A Diffusion Hydrodynamic Model

Authored by Theodore V. Hromadka II, Chung-Cheng Yen and Prasada Rao

Published in London, United Kingdom

IntechOpen

Supporting open minds since 2005

A Diffusion Hydrodynamic Model
http://dx.doi.org/10.5772/intechopen.90224
Authored by Theodore V. Hromadka II, Chung-Cheng Yen and Prasada Rao

First published in London, United Kingdom, 2020 by IntechOpen
IntechOpen is the global imprint of INTECHOPEN LIMITED, registered in England and Wales, registration number: 11086078, 7th floor, 10 Lower Thames Street, London, EC3R 6AF, United Kingdom
Printed in Croatia

British Library Cataloguing-in-Publication Data
A catalogue record for this book is available from the British Library

Additional hard and PDF copies can be obtained from orders@intechopen.com

A Diffusion Hydrodynamic Model
Authored by Theodore V. Hromadka II, Chung-Cheng Yen and Prasada Rao
p. cm.
Print ISBN 978-1-83962-817-7
Online ISBN 978-1-83962-818-4
eBook (PDF) ISBN 978-1-83962-819-1

An electronic version of this book is freely available, thanks to the support of libraries working with Knowledge Unlatched. KU is a collaborative initiative designed to make high quality books Open Access for the public good. More information about the initiative and links to the Open Access version can be found at www.knowledgeunlatched.org

We are IntechOpen,
the world's leading publisher of
Open Access books
Built by scientists, for scientists

5,000+
Open access books available

125,000+
International authors and editors

140M+
Downloads

Our authors are among the

151
Countries delivered to

Top 1%
most cited scientists

12.2%
Contributors from top 500 universities

CLARIVATE ANALYTICS
BOOK
CITATION
INDEX
INDEXED

WEB OF SCIENCE™

Selection of our books indexed in the Book Citation Index
in Web of Science™ Core Collection (BKCI)

Interested in publishing with us?
Contact book.department@intechopen.com

Numbers displayed above are based on latest data collected.
For more information visit www.intechopen.com

Meet the authors

Hromadka & Associates' Principal and Founder, Theodore Hromadka II, PhD, PhD, PhD, PH, PE, has extensive scientific, engineering, expert witness, and litigation support experience. His frequently referenced scientific contributions to the hydrologic, earth, and atmospheric sciences have been widely published in peer-reviewed scientific literature, including 30 books and more than 500 scientific papers, book chapters, and government reports. His professional engineering experience includes supervision and development of over 1500 engineering studies. He is currently a faculty member at the United States Military Academy at West Point, New York.

Chung-Cheng Yen received his Ph.D. degree from the University of California, Irvine, in 1985. He has more than 35 years of experience in the field of water resource engineering, specializing in hydrology, hydraulics, dam breach, and groundwater modeling. His work experience includes rainfall analysis, flood frequency analysis, rainfall-runoff modeling, detention basin flood routing analysis, drainage master plan, FEMA floodplain evaluations and mapping, dam breach analysis and flood inundation mapping, and the USACE risk and uncertainty analysis. Dr. Yen has conducted floodplain analyses using 2-D hydrodynamic models (such as DHM, FLO-2D, HEC-RAS 1D/2D, and XPSWMM), prepared hydrologic and hydraulic studies for government and private entities, and drainage master plans for various cities in southern California.

Prasada Rao is a Professor in the Civil and Environmental Engineering Department at California State University, Fullerton. His current research areas relate to surface and subsurface flow modeling and computational mathematics. He has worked extensively on developing innovative, hydraulic and hydrological modelling solutions to better predict surface flow phenomena along with its impact on groundwater levels. He has also worked on developing parallel hydraulic models for large scale applications. He has taught undergraduate and graduate level courses in hydraulics, hydrology, open channel flow, and hydraulic structures.

Contents

Preface

The Diffusion Hydrodynamic Model (DHM), as presented in the 1987 USGS publication (https://pubs.er.usgs.gov/publication/wri874137), was one of the first computational fluid dynamics computational programs based on the groundwater program MODFLOW, which evolved into the control volume modeling approach. In the DHM, overland flow effects are modeled by a two-dimensional unsteady flow hydraulic model based on the diffusion (non-inertial) form of the governing flow equations. The channel flow is modeled using a one-dimensional unsteady flow hydraulic model based on the diffusion type equation. DHM can simulate both approximate unsteady supercritical and subcritical flow (without the user predetermining hydraulic controls), backwater flooding effects, and escaping and returning flow from the two-dimensional overland flow model to the channel system. The model is also capable of treating such effects as backwater, drawdown, channel overflow, storage, and ponding.

Since 1987, others developed similar computational programs that either used the methodology and approaches presented in the DHM directly or were its extensions that included additional components and capacities. Later, the DHM itself was extended considerably to the version EDHM (Extended DHM), although the fundamental mechanics of the procedures were retained.

The original effort was funded by the USGS, and the authors acknowledge their support. The report submitted to the USGS is available online at https://pubs. er.usgs.gov/publication/wri874137, and some of the relevant contract details are:

Water Resources Investigations Report: 87- 4137

Name of Contractor: Williamson and Schmid

Principal Investigator: Theodore V. Hromadka II

Contract Officer's Representative: Marshall E. Jennings

Short Title of Work: Diffusion Hydrodynamic Model

Year Published: 1987

The time evolution of this document from the original 1987 USGS report to the present content in this book is summarized below.

Although the original report is available on the web as a pdf file, we were unable to locate the relevant computer files on our memory devices. Correspondence with the USGS also pointed only to the report that is on the web and not to any electronic files available offline. Since some of the pages and figures in the pdf report lacked clarity, we started this book by retyping the entire report (as it is) along with the equations using MS Word. A few graduate students from the Civil and Environmental Engineering Department at California State University, Fullerton did the retyping task, and we acknowledge their effort. While many figures were also redrawn, some of the figures (because of the complexity) were left as they were presented in the original report.

Our goal is to show the readers that the Diffusion Hydrodynamic Model, which was developed in an age preceding computer graphics/visualization tools, is as robust as any of the popular models that are currently used in the consulting industry. To this end, we wanted to enhance/revise the original report by adding a new chapter that compares the results of the DHM with current standard models, including HEC-RAS, TUFLOW, Mike 21, RAS 2D, WSPG, and OpenFOAM applied to a few complex flows and physical domain scenarios. Since we were building on the original USGS report, approval from the USGS was obtained to enhance the original report. We thank the USGS for their approval and for permitting us to use the content from the earlier USGS report.

Specific major additions/deletions to the text in the original report are:

(1) The DHM Fortran source code was deleted. Since the source code (DHM21. FOR) and its executable file for Windows environment (DHM21.EXE), along with the executable file for extended DHM (EDHM21.EXE) and the sample data file can be downloaded from *www.diffusionhydrodynamicmodel.com,* we did not see an advantage for again listing the source code and the data file.

(2) Chapter 6 has been added.

Minor formatting changes were made to the content in the original report to make it compatible with the publisher's guidelines. We hope that this report, together with the resources present at the companion website, *http://diffusionhydrodynamicmodel. com,* will motivate the readers to use DHM for their applications. The resources in the companion website include:

- DHM Program source code (DHM21.FOR) and its executable code (DHM21. EXE).

- Executable code for the extended DHM (EDHM21.EXE).

- Sample input data files and related publications/presentations.

In this book, ample applications of DHM are included, which hopefully demonstrate the utility of this modeling approach in many drainage engineering problems. The model is applied to a collection of one- and two-dimensional unsteady flows hydraulic problems including (1) one-dimensional unsteady flow problem, (2) rainfall-runoff model, (3) dam-break flow analysis, (4) estuary model, (5) channel floodplain interface model, (6) mixed flows in open channel, (7) overland flow, and (8) flow through a constriction. For selected applications, DHM results have been compared with those from other widely used hydraulic and CFD models. Consequently, the diffusion hydrodynamic model promises to result in a highly useful, accurate, and simple to use computer model, which is of immediate use to practicing flood control engineers. Use of the DHM in surface runoff problems will result in a highly versatile and practical tool which significantly advances the current state-of-the-art flood control system and flood plain mapping analysis procedures, resulting in more accurate predictions in the needs of the flood control system, and potentially proving a considerable cost saving due to reduction of conservation used to compensate for the lack of proper hydraulic unsteady flow effects approximation.

Theodore V. Hromadka II
Department of Mathematical Sciences,
United States Military Academy,
West Point, NY, USA

Chung-Cheng Yen
Tetra Tech,
Irvine, CA, USA

Prasada Rao
Department of Civil and Environmental Engineering,
California State University,
Fullerton, CA, USA

Chapter 1

Diffusion Hydrodynamic Model Theoretical Development

Theodore V. Hromadka II and Chung-Cheng Yen

Abstract

In this chapter, the governing flow equations for one- and two-dimensional unsteady flows that are solved in the diffusion hydrodynamic model (DHM) are presented along with the relevant assumptions. A step-by-step derivation of the simplified equations which are based on continuity and momentum principles are detailed. Characteristic features of the explicit DHM numerical algorithm are discussed.

Keywords: unsteady flow, conservation of mass, finite difference, explicit scheme, flow equations

1. Introduction

Many flow phenomena of great engineering importance are unsteady in characters and cannot be reduced to a steady flow by changing the viewpoint of the observer. A complete theory of unsteady flow is therefore required and will be reviewed in this section. The equations of motion are not solvable in the most general case, but approximations and numerical methods can be developed which yield solutions of satisfactory accuracy.

2. Review of governing equations

The law of continuity for unsteady flow may be established by considering the conservation of mass in an infinitesimal space between two channel sections (**Figure 1**). In unsteady flow, the discharge, Q, changes with distance, x, at a rate $\frac{\partial Q}{\partial x}$, and the depth, y, changes with time, t, at a rate $\frac{\partial y}{\partial t}$. The change in discharge volume through space dx in the time dt is $\left(\frac{\partial Q}{\partial x}\right)dxdt$. The corresponding change in channel storage in space is $Tdx\left(\frac{\partial y}{\partial t}\right)dt = dx\left(\frac{\partial A}{\partial t}\right)dt$ in which $A = Ty$. Because water is incompressible, the net change in discharge plus the change in storage should be zero, that is

$$\left(\frac{\partial Q}{\partial x}\right)dxdt + Tdx\left(\frac{\partial y}{\partial t}\right)dt = \left(\frac{\partial Q}{\partial x}\right)dxdt + dx\left(\frac{\partial A}{\partial t}\right)dt = 0$$

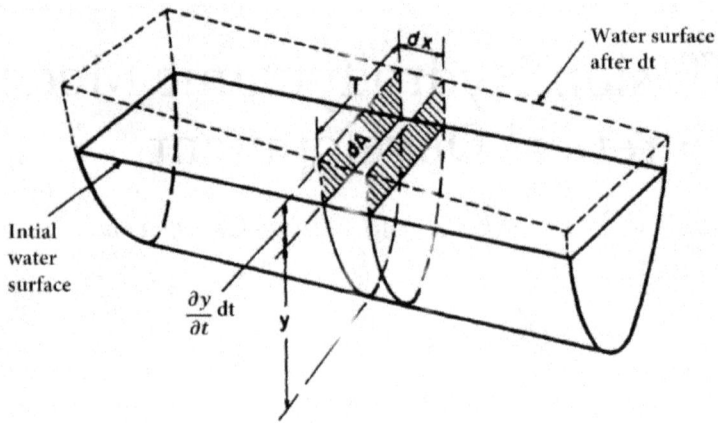

Figure 1.
Continuity of unsteady flow.

Simplifying

$$\frac{\partial Q}{\partial x} + T\frac{\partial y}{\partial t} = 0 \tag{1}$$

or

$$\frac{\partial Q}{\partial x} + \frac{\partial A}{\partial t} = 0 \tag{2}$$

At a given section, $Q = VA$; thus Eq. (1) becomes

$$\frac{\partial(VA)}{\partial x} + T\frac{\partial y}{\partial t} = 0 \tag{3}$$

or

$$A\frac{\partial V}{\partial x} + V\frac{\partial A}{\partial x} + T\frac{\partial y}{\partial t} = 0 \tag{4}$$

Because the hydraulic depth $D = A/T$ and $\partial A = T\partial y$, the above equation may be written as

$$D\frac{\partial V}{\partial x} + V\frac{\partial y}{\partial x} + \frac{\partial y}{\partial t} = 0 \tag{5}$$

The above equations are all forms of the continuity equation for unsteady flow in open channels. For a rectangular channel or a channel of infinite width, Eq. (1) may be written as

$$\frac{\partial q}{\partial x} + \frac{\partial y}{\partial t} = 0 \tag{6}$$

where q is the discharge per unit width.

3. Equation of motion

In a steady, uniform flow, the gradient, $\frac{dH}{dx}$, of the total energy line is equal to magnitude of the "friction slope" $S_f = V^2/(C^2 R)$, where C is the Chezy coefficient and R is the hydraulic radius. Indeed this statement was in a sense taken as the definition of S_f; however, in the present context, we have to consider the more general case in which the flow is nonuniform, and the velocity may be changing in the downstream direction. The net force, shear force and pressure force, is no longer zero since the flow is accelerating. Therefore, the equation of motion becomes

$$-\gamma A \Delta h - \tau_0 P \Delta x = \rho A \Delta x \left(V \frac{\partial V}{\partial x} + \frac{\partial V}{\partial t} \right)$$

that is

$$\tau_0 = -\gamma R \left(\frac{\partial h}{\partial x} + \frac{V}{g} \frac{\partial V}{\partial x} + \frac{1}{g} \frac{\partial V}{\partial t} \right)$$

$$-\gamma R \left(\frac{\partial H}{\partial x} + \frac{1}{g} \frac{\partial V}{\partial t} \right) \tag{7}$$

where τ_0 is the same shear stress, P is the hydrostatic pressure, h is the depth of water, Δh is the change of depth of water, γ is the specific weight of the fluid, R is the mean hydraulic radius, and ρ is the fluid density. Substituting $\frac{\tau_0}{\gamma R} = \frac{V^2}{C^2 R}$ into Eq. (7), we obtain

$$\frac{\partial H}{\partial x} + \frac{1}{g} \frac{\partial V}{\partial t} + \frac{V^2}{C^2 R} = 0 \tag{8}$$

and this equation may be rewritten as

$$S_e + S_a + S_f = 0 \tag{9}$$

where the three terms of Eq. (9) are called the energy slope, the acceleration slope, and the friction slope, respectively. **Figure 2** depicts the simplified representation of energy in unsteady flow.

By substituting $H = \frac{V^2}{2g} + y + z$ and the bed slope $S_o = -\frac{dz}{dx}$ into Eq. (8), we obtain

$$\frac{\partial H}{\partial x} = \frac{\partial z}{\partial x} + \frac{\partial y}{\partial x} + \frac{V}{g} \frac{\partial V}{\partial x}$$

$$= -S_o + \frac{\partial y}{\partial x} + \frac{V}{g} \frac{\partial V}{\partial x} \tag{10}$$

$$= -\frac{1}{g} \frac{\partial V}{\partial t} - S_f$$

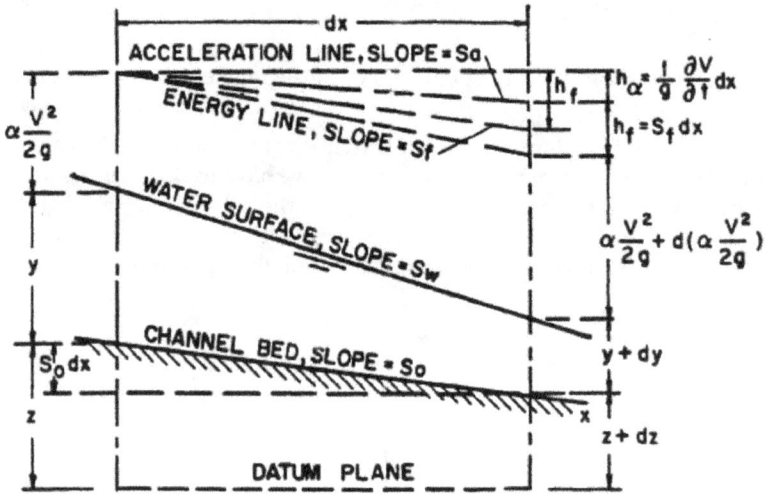

Figure 2.
Simplified representation of energy in unsteady flow.

Hence Eq. (8) can be rewritten as

$$S_f = S_0 - \frac{\partial y}{\partial x} - \frac{V}{g}\frac{\partial V}{\partial x} - \frac{1}{g}\frac{\partial V}{\partial t} = \frac{V^2}{C^2 R}$$

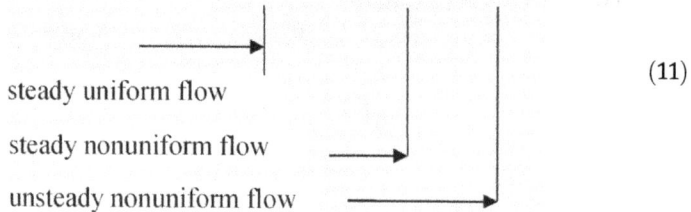

steady uniform flow

steady nonuniform flow

unsteady nonuniform flow

(11)

This equation may be applicable to various types of flow as indicated. This arrangement shows how the nonuniformity and unsteadiness of flows introduce extra terms into the governing dynamic equation.

4. Diffusion hydrodynamic model

4.1 One-dimensional diffusion hydrodynamic model

The mathematical relationships in a one-dimensional diffusion hydrodynamic model (DHM) are based upon the flow equations of continuity (2) and momentum (11) which can be rewritten [1] as

$$\frac{\partial Q_x}{\partial x} + \frac{\partial A_x}{\partial t} = 0 \tag{12}$$

$$\frac{\partial Q_x}{\partial t} + \frac{\partial \left(Q_x{}^2 / A_x\right)}{\partial x} + g A_x \left(\frac{\partial H}{\partial x} + S_{fx}\right) = 0 \tag{13}$$

where Q_x is the flow rate; x,t are spatial and temporal coordinates, A_x is the flow area, g is the gravitational acceleration, H is the water surface elevation, and S_{fx} is a friction slope. It is assumed that S_{fx} approximated from Manning's equation for steady flow by [1].

$$Q_x = \frac{1.486}{n} A_x R^{2/3} S_{fx}^{1/2} \tag{14}$$

where R is the hydraulic radius and n is a flow resistance coefficient which may be increased to account for other energy losses such as expansions and bend losses. Letting m_x be a momentum quantity defined by

$$m_x = \left(\frac{\partial Q_x}{\partial t} + \frac{\partial(Q_x^2/A_x)}{\partial x}\right)/gA_x \tag{15}$$

then Eq. (13) can be rewritten as

$$S_{fx} = -\left(\frac{\partial H}{\partial x} + m_x\right) \tag{16}$$

In Eq. (15), the subscript x included in m_x indicates the directional term. The expansion of Eq. (13) to two-dimensional case leads directly to the terms (m_x, m_y) except that now a cross-product of flow velocities is included, increasing the computational effort considerably.

Rewriting Eq. (14) and including Eqs. (15) and (16), the directional flow rate is computed by

$$Q_x = -K_x\left(\frac{\partial H}{\partial x} + m_x\right) \tag{17}$$

where Q_x indicates a directional term and K_x is a type of conduction parameter defined by

$$K_x = \frac{1.486}{n} \frac{A_x R^{2/3}}{\left|\frac{\partial H}{\partial x} + m_x\right|^{1/2}} \tag{18}$$

In Eq. (18), K_x is limited in value by the denominator term being checked for a smallest allowable magnitude (such as $\left|\frac{\partial H}{\partial X} + m_X\right|^{1/2} > 10^{-3}$).

Substituting the flow rate formulation of Eq. (17) into Eq. (12) gives a diffusion type of relationship

$$\frac{\partial}{\partial X} K_X\left[\frac{\partial H}{\partial X} + m_X\right] = \frac{\partial A_X}{\partial t} \tag{19}$$

The one-dimensional model of Akan and Yen [1] assumed $m_X = 0$ in Eq. (18). The m_X term is assumed to be negligible when combined with the other similar terms—that is, they are considered as a sum rather than as individual directional terms that typically have more significance when examined individually. Additionally, the term "diffusion" routing indicates assuming that several convective and other components have a small contribution to the coupled mass and energy balance equations and therefore are neglected in the computational formulation to simplify the model accordingly. Thus, the one-dimensional DHM equation is given by

$$\frac{\partial}{\partial X} K_X \frac{\partial H}{\partial X} = \frac{\partial A_X}{\partial t} \tag{20}$$

where K_X is now simplified as

$$K_x = \frac{\frac{1.486}{n} A_x R^{2/3}}{\left|\frac{\partial H}{\partial X}\right|^{\frac{1}{2}}} \tag{21}$$

For a channel of constant width, W_X, Eq. (20) reduces to

$$\frac{\partial}{\partial X} K_X \frac{\partial H}{\partial X} = W_X \frac{\partial H}{\partial t} \tag{22}$$

Assumptions other than $m_X = 0$ in Eq. (19) result in a family of models:

$$m_x = \begin{cases} \dfrac{\partial(Q_x{}^2/A_X)}{\partial X} \Big/ gA_X & \text{(convective acceleration model)} \\[2ex] \dfrac{\partial Q_X}{\partial t} \Big/ gA_X & \text{(local acceleration model)} \\[2ex] \left[\dfrac{\partial Q_X}{\partial t} + \dfrac{\partial(Q_x{}^2/A_X)}{\partial X}\right] \Big/ gA_X & \text{(fully dynamic model)} \\[2ex] 0 & \text{(DHM)} \end{cases} \tag{23}$$

4.2 Two-dimensional diffusion hydrodynamic model

The set of (fully dynamic) 2D unsteady flow equations consists of one equation of continuity

$$\frac{\partial q_x}{\partial x} + \frac{\partial q_y}{\partial y} + \frac{\partial H}{\partial t} = 0 \tag{24}$$

and two equations of motion

$$\frac{\partial q_x}{\partial t} + \frac{\partial}{\partial x}\left[\frac{q_x{}^2}{h}\right] + \frac{\partial}{\partial y}\left[\frac{q_x q_y}{h}\right] + gh\left[S_{fx} + \frac{\partial H}{\partial X}\right] = 0 \tag{25}$$

$$\frac{\partial q_y}{\partial t} + \frac{\partial}{\partial y}\left[\frac{q_y{}^2}{h}\right] + \frac{\partial}{\partial y}\left[\frac{q_x q_y}{h}\right] + gh\left[S_{fy} + \frac{\partial H}{\partial y}\right] = 0 \tag{26}$$

where q_X and q_y are flow rates per unit width in the x and y directions; S_{fx} and S_{fy} represents friction slopes in x and y directions; H, h, and g stand for water surface elevation, flow depth, and gravitational acceleration, respectively; and x, y, and t are spatial and temporal coordinates.

The above equation set is based on the assumptions of constant fluid density without sources or sinks in the flow field and of hydrostatic pressure distributions.

The local and convective acceleration terms can be grouped, and Eqs. (25) and (26) are rewritten as

$$m_z + \left[S_{fz} + \frac{\partial H}{\partial Z}\right] = 0, z = x, y \tag{27}$$

where m_Z represents the sum of the first three terms in Eqs. (25) or (26) divided by gh. Assuming the friction slope to be approximated by the Manning's formula, one obtains, in the US customary units for flow in the x or y directions,

$$q_z = \frac{1.486}{n} h^{5/3} S_{fz}^{1/2}, z = x, y \tag{28}$$

Eq. (28) can be rewritten in the general case as

$$q_z = -K_z \frac{\partial H}{\partial Z} - K_z m_z, z = x, y \tag{29}$$

where

$$K_z = \frac{1.486}{n} \frac{h^{5/3}}{\left| \frac{\partial H}{\partial S} + m_S \right|^{1/2}}, z = x, y \tag{30}$$

The symbol s in Eq. (30) indicates the flow direction which makes an angle of $\theta = tan^{-1}\left(q_y/q_x\right)$ with the positive x direction.

The m_z term is assumed to be negligible [1–5] when combined with the other similar terms, i.e., they are considered as a sum rather than as individual directional terms that typically have more significance when examined individually. Additionally, the term "diffusion" routing indicates assuming that several convective and other components have a small contribution to the coupled mass and energy balance equations and therefore are neglected in the computational formulation to simplify the model accordingly. Neglecting this term results in the simple diffusion model

$$q_z = -K_z \frac{\partial H}{\partial Z}, z = x, y \tag{31}$$

The proposed 2D DHM is formulated by substituting Eq. (31) into Eq. (24)

$$\frac{\partial}{\partial X} K_x \frac{\partial H}{\partial X} + \frac{\partial}{\partial y} K_y \frac{\partial H}{\partial y} = \frac{\partial H}{\partial t} \tag{32}$$

If the momentum term groupings were retained, Eq. (32) would be written as

$$\frac{\partial}{\partial x} K_x \frac{\partial H}{\partial x} + \frac{\partial}{\partial y} K_y \frac{\partial H}{\partial y} + S = \frac{\partial H}{\partial t} \tag{33}$$

where

$$S = \frac{\partial}{\partial x} (K_x m_x) + \frac{\partial}{\partial y} (K_x m_y)$$

and K_x and K_y are also functions of m_x and m_y, respectively.

5. Numerical approximation

5.1 Numerical solution algorithm

The one-dimensional domain is discretized across uniformly spaced nodal points, and at each of these points, at time (t) = 0, the values of Manning's n, an elevation, and initial flow depth (usually zero) are assigned. With these initial conditions, the solution is advanced to the next time step (t + Δt) as detailed below

1. Between nodal points, compute an average Manning's n and average geometric factors

2. Assuming $m_X = 0$, estimate the nodal flow depths for the next time step (t + Δt) by using Eqs. (20) and (21) explicitly

3. Using the flow depths at time t and t + Δt, estimate the mid time step value of m_X selected from Eq. (23)

4. Recalculate the conductivities K_X using the appropriate m_X values

5. Determine the new nodal flow depths at the time (t + Δt) using Eq. (19), and

6. Return to step (3) until K_X matches mid time step estimates.

The above algorithm steps can be used regardless of the choice of definition for m_X from Eq. (23). Additionally, the above program steps can be directly applied to a two-dimensional diffusion model with the selected (m_X, m_y) relations incorporated.

5.2 Numerical model formulation (grid element)

For uniform grid elements, the integrated finite difference version of the nodal domain integration (NDI) method [6] is used. For grid elements, the NDI nodal equation is based on the usual nodal system shown in **Figure 3**. Flow rates across the boundary Γ are estimated by assuming a linear trial function between nodal points. For a square grid of width δ

$$q|\Gamma_E = -[K_X|\Gamma_E]\,[H_E - H_C]/\delta \tag{34}$$

where

$$K_{x|\Gamma_E} \begin{cases} \left[\dfrac{1.486}{n}h^{5/3}\right]_{\Gamma_E} \bigg/ \left|\dfrac{H_E - H_C}{\delta \cos\theta}\right|^{1/2} & ; |H_E - H_c| \geq \varepsilon \\ 0 & ; |H_E - H_C| < \varepsilon \end{cases} \tag{35}$$

In Eq. (35), h (depth of water) and n (the Manning's coefficient) are both the average of their respective values at C and E, i.e., $h = (h_C + h_E)/2$ and $n = (n_C + n_E)/2$. Additionally, the denominator of K_X is checked such that K_X is set to zero if $|H_E - H_C|$ is less than a tolerance ε such as 10^{-3} ft.

The net volume of water in each grid element between time step i and i + 1 is $\Delta q_C{}^i = q|r_E + q|r_w + q|r_N + q|r_S$ and the change of depth of water is $\Delta H_C{}^i = \Delta q_C{}^i * \Delta t/\delta^2$ for time step i and i + 1 with Δt interval. Then the model advances in time by an explicit approach

$$H_C{}^{i+1} = \Delta H_C{}^i + H_C{}^i \tag{36}$$

where the assumed input flood flows are added to the specified input nodes at each time step. After each time step, the hydraulic conductivity parameters of Eq. (35) are reevaluated, and the solution of Eq. (36) is reinitiated.

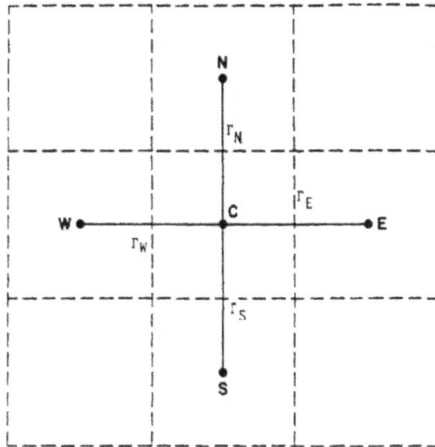

Figure 3.
Two-dimensional finite difference analog.

5.3 Model time step selection

The sensitivity of the model to time step selection is dependent upon the slope of the discharge hydrograph $(\frac{\partial Q}{\partial t})$ and the grid spacing. Increasing the grid spacing size introduces additional water storage to a corresponding increase in nodal point flood depth values. Similarly, a decrease in time step size allows a refined calculation of inflow and outflow values and a smoother variation in nodal point flood depths with respect to time. The computer algorithm may self-select a time step by increments of halving (or doubling) the initial user-chosen time step size so that a proper balance of inflow-outflow to control volume storage variation is achieved. In order to avoid a matrix solution for flood depths, an explicit time stepping algorithm is used to solve for the time derivative term. For large time steps or a rapid variation in the dam-break hydrograph (such as $\frac{\partial Q}{\partial t}$ is large), a large accumulation of flow volume will occur at the most upstream nodal point. That is, at the dam-break reservoir nodal point, the lag in outflow from the control volume can cause an unacceptable error in the computation of the flood depth. One method that offsets this error is the program to self-select the time step until the difference in the rate of volume accumulation is within a specified tolerance.

Due to the form of the DHM in Eq. (22), the model can be extended into an implicit technique. However, this extension would require a matrix solution process which may become unmanageable for two-dimensional models which utilize hundreds of nodal points.

6. Conclusions

The one- and two-dimensional flow equations used in the diffusion hydrodynamic model are derived, and the relevant assumptions are listed. These equations, which are the basis of the model, are based on the conservation of mass and momentum principles. The explicit numerical algorithm and the discretized equations are also presented. The ability of the model to self-select the optimal time step is discussed.

Author details

Theodore V. Hromadka II[1]* and Chung-Cheng Yen[2]

1 Department of Mathematical Sciences, United States Military Academy, West Point, NY, USA

2 Tetra Tech, Irvine, CA, USA

*Address all correspondence to: tedhromadka@yahoo.com

IntechOpen

References

[1] Akan AO, Yen BC. Diffusion-wave flood routing in channel networks. ASCE Journal of Hydraulic Division. 1981;**107**(6):719-732

[2] Hromadka TV II, Berenbrokc CE, Freckleton JR, Guymon GL. A two-dimensional diffusion dam-break model. Advances in Water Resources. 1985;**8**:7-14

[3] Xanthopoulos TH, Koutitas CH. Numerical simulation of a two-dimensional flood wave propagation due to dam failure. ASCE Journal of Hydraulic Research. 1976;**14**(4):321-331

[4] Hromadka TV II, Lai C. Solving the two-dimensional diffusion flow model. In: Proceedings of ASCE Hydraulics Division Specialty Conference, Orlando, Florida. 1985

[5] Hromadka TV II, Nestlinger AB. Using a two-dimensional diffusional dam-break model in engineering planning. In: Proceedings of ASCE Workshop on Urban Hydrology and Stormwater Management, Los Angeles County Flood Control District Office, Los Angeles, California. 1985

[6] Hromadka TV II, Guymon GL, Pardoen G. Nodal domain integration model of unsaturated two-dimensional soil water flow: Development. Water Resources Research. 1981;**17**:1425-1430

Chapter 2

Verification of Diffusion Hydrodynamic Model

Theodore V. Hromadka II and Chung-Cheng Yen

Abstract

The efficacy of the one- and two-dimensional diffusion hydrodynamic model (DHM) for predicting flow characteristics resulting from a dam-break scenario is tested. The model results, for different inflow scenarios, are compared with the standard United States Geological Survey (USGS) K-634 model. The sensitivity of the model results to grid spacing and the chosen time step are presented. The model results are in close agreement.

Keywords: floodplain, hydrograph, unsteady flows, initial flow condition, spatial grid size, transient simulation

1. Introduction

An unsteady flow hydraulic problem of considerable interest is the analysis of dam breaks and their downstream hydrograph. In this section, the main objective is to evaluate the diffusion form of the flow equations for the estimation of flood depths (and the floodplain), resulting from a specified dam-break hydrograph. The dam-break failure mode is not considered in this section. Rather, the dam-break failure mode may be included as part of the model solution (such as for a sudden breach) or specified as a reservoir outflow hydrograph.

The use of numerical methods to approximately solve the flow equations for the propagation of a flood wave due to an earthen dam failure has been the subject of several studies reported in the literature. Generally, the flow is modeled using the one-dimensional equation wherever there is no significant lateral variation in the flow. Land [1, 2] examined four such dam-break models in his prediction of flooding levels and flood wave travel time and compares the results against observed dam failure information. In the dam-break analysis, an assumed dam-break failure mode (which may be part of the solution) is used to develop an inflow hydrograph to the downstream floodplain. Consequently, it is noted that a considerable sensitivity in modeling results is attributed to the dam-break failure rate assumptions. Ponce and Tsivoglou [3] examined the gradual failure of an earthen embankment (caused by an overtopping flooding event) and present detailed analysis for each part of the total system: sediment transport, unsteady channel hydraulics, and earth embankment failure.

In another study, Rajar [4] studied a one-dimensional flood wave propagation from an earthen dam failure. His model solved the St. Venant equations using either a first-order diffusive or a second-order Lax-Wendroff numerical scheme. A review of the literature indicates that the most frequently used numerical scheme was the method of characteristics (to solve the governing flow equations) as described in Sakkas and Strelkoff [5], Chen [6], and Chen and Armbruster [7].

Although many dam-break studies involve flood flow regimes which are truly two-dimensional (in the horizontal plane), the two-dimensional case has not received much attention in the literature. Katopodes and Strelkoff [8] used the method of bicharacteristics to solve the governing equations of continuity and momentum. The model utilizes a moving grid algorithm to follow the flood wave propagation and also employs several interpolation schemes to approximate the nonlinearity effects. In a much simpler approach, Xanthopoulos and Koutitas [9] used a diffusion model (i.e., the inertial terms are assumed negligible in comparison to the pressure, friction, and gravity components) to approximate a two-dimensional flow field. The model assumed that the flow regime in the floodplain is such that the inertial terms (local and convective acceleration) are negligible. In a one-dimensional model, Akan and Yen [10] also used the diffusion approach to model hydrograph confluences at channel junctions. In the latter study, comparisons of modeling results were made between the diffusion model, a complete dynamic wave model solving the total equation system, and the basic kinematic-wave equation model (i.e., the inertia and pressure terms are assumed negligible in comparison to the friction and gravity terms). The differences between the diffusion model and the dynamic wave model were small, showing only minor discrepancies.

The kinematic-wave flow model has been used in the computation of dam-break flood waves [11]. Hunt concluded in his study that the kinematic-wave solution is asymptotically valid. Since the diffusion model has a wider range of applicability for varied bed slopes and wave periods than the kinematic model [12], the diffusion model approach should provide an extension to the referenced kinematic model.

Because the diffusion modeling approach leads to an economic two-dimensional dam-break flow model (with numerical solutions based on the usual integrated finite difference or finite element techniques), the need to include the extra components in the momentum equation must be ascertained. For example, evaluating the convective acceleration terms in a two-dimensional flow model requires approximately an additional 50 percent of the computational effort required in solving the entire two-dimensional model with the inertial terms omitted. Consequently, including the local and convective acceleration terms increases the computer execution costs significantly. Such increases in computational effort may not be significant for one-dimensional case studies; however, two-dimensional case studies necessarily involve considerably more computational effort, and any justifiable simplifications of the governing flow equations is reflected by a significant decrease in computer software requirements, costs, and computer execution time.

Ponce [13] examined the mathematical expressions of the flow equations, which lead to wave attenuation in prismatic channels. It is concluded that the wave attenuation process is caused by the interaction of the local acceleration term with the sum of the terms of friction slope and channel slope. When local acceleration is considered negligible, wave attenuation is caused by the interaction of the friction slope and channel slope terms with the pressure gradient or convective acceleration terms (or a combination of both terms). Other discussions of flow conditions and the sensitivity to the various terms of the flow equations are given in Miller and Cunge [14], Morris and Woolhiser [15], and Henderson [16].

It is stressed that the ultimate objective of this effort is to develop a two-dimensional diffusion model for use in estimating floodplain evolution, such as those that occur due to drainage system deficiencies. Prior to finalizing such a model, the requirement of including the inertial terms in the unsteady flow equations needs to be ascertained. The strategy used to check on this requirement is to evaluate the accuracy in predicted flood depths produced from a one-dimensional diffusion model with respect to the one-dimensional United States Geological Survey (USGS) K-634 dam-break model which includes all of the inertial term components.

2. One-dimensional analysis

2.1 Study approach

To evaluate the accuracy of the one-dimensional diffusion model [Chapter 1, Eq. 22) in the prediction of flood depths, the USGS fully dynamic flow model K-634 [1, 2] is used to determine channel flood depths for comparison purposes. The K-634 model solves the coupled flow equations of continuity and momentum by an implicit finite difference approach and is considered to be a highly accurate model for many unsteady flow problems. The study approach is to compare predicted (1) flood depths and (2) discharge hydrographs from both the K-634 and the diffusion hydrodynamic model (DHM) for various channel slopes and inflow hydrographs.

It should be noted that different initial conditions are used for these two models. The USGS K-634 model requires a base flow to start the simulation; therefore, the initial depth of water cannot be zero. Next, the normal depth assumption is used to generate an initial water depth before the simulation starts. These two steps are not required by the DHM.

In this case study, two hydrographs are assumed; namely, peak flows to 120,000 and 600,000 cfs. A base flow of 5000 and 40,000 cfs was used for hydrographs with peaks of 120,000 and 600,000 cfs, respectively, for all K-634 simulations. Both hydrographs are assumed to increase linearly from zero (or the base flow) to the peak flow rate at 1 h and then decrease linearly to zero (or the base flow) at 6 h (see **Figure 1** inset). The study channel is assumed to be a 1000-feet-width rectangular section of Manning's n equal to 0.040 and various slopes S_0 in the range of $0.001 \leq S_0 \leq 0.01$. **Figure 1** shows the comparison of modeling results. From the figure, various flood depths are plotted along the channel length of up to 10 miles. Two reaches of channel lengths of up to 30 miles are also plotted in **Figure 1** which correspond to a slope $S_0 = 0.0020$. In all tests, grid spacing was set at 1000-feet intervals. Time steps were 0.01 h for K-634 and 7.2 s for DHM.

From **Figure 1**, it is seen that the diffusion model provides estimates of flood depths that compare very well to the flood depths predicted from the K-634 model. For downstream distances at up to 30 miles, differences in predicted flood depths are less than 3% for the various channel slopes and peak flow rates considered.

In **Figures 2–5**, good comparisons between the diffusion hydrodynamic and the K-634 models are observed for water depths and outflow hydrographs at 5 and 10 miles downstream from the dam-break site. It should be noted that the test conditions are purposefully severe in order to bring out potential inaccuracies in the diffusion hydrodynamic model results. Less severe test conditions should lead to more favorable comparisons between the two model results. Although offsets do occur in timing, volume continuity is preserved when allowances are made for differences in base flow volumes.

2.2 Grid spacing selection

The choice of the time step and grid size for an explicit time advancement is a relative matter and is theoretically based on the well-known Courant condition [17]. The choice of grid size usually depends on available topographic data for nodal elevation determination and the size of the problem. The effect of the grid size

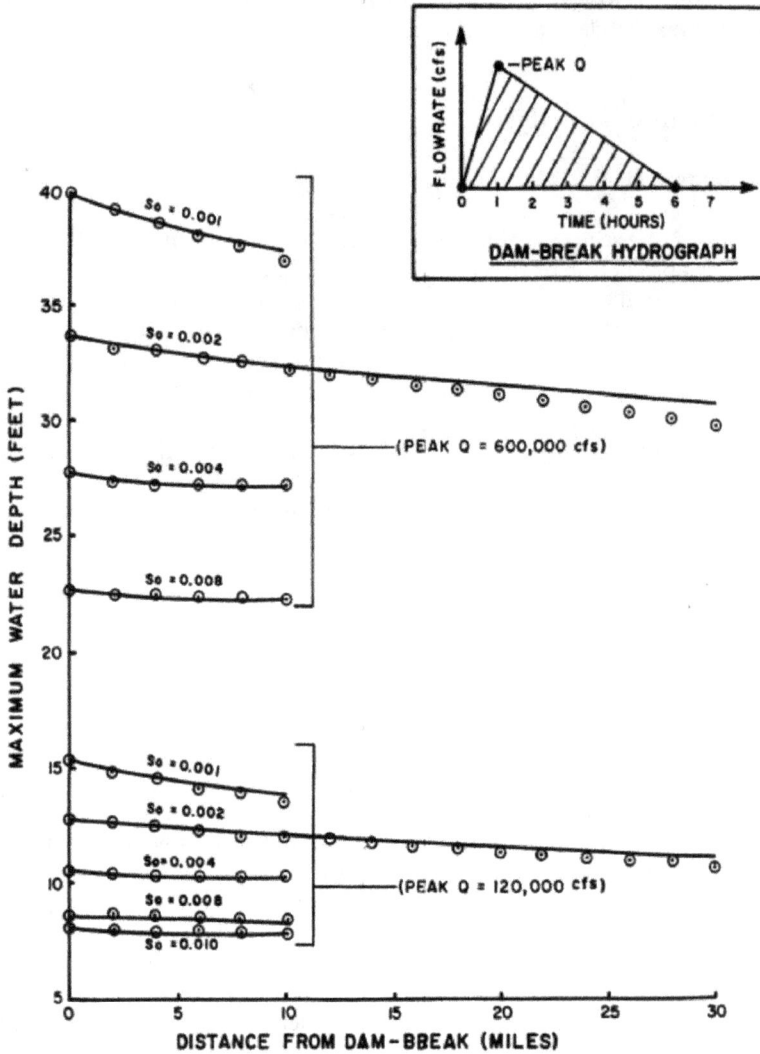

Figure 1.
Diffusion hydrodynamic model (⊙) and K-634 model results (solid line) for 1000-feet-width channel, Manning's n = 0.040, and various channel slopes, S_0.

(for constant time step for 7.2 s) on the diffusion model accuracy can be shown by example where nodal spacings of 1000, 2000, and 5000 feet are considered. The predicted flood depths varied only slightly from choosing the grid size between 1000 and 2000 feet. However, an increased variation in results occurs when a grid size of 5000 feet is selected. For the example of peak flow rate test hydrograph of 600,000 cfs, the differences of simulated flow depths between 1000 and 5000-feet grid are 0.03, 0.06, and 0.17 feet at 1, 5, and 10 miles, respectively, downstream from the dam-break site for the maximum flow depth with the magnitude of 30 feet.

Because the algorithm presented is based upon an explicit time stepping technique, the modeling results may become inaccurate, should the time step size versus grid size ratio become large. A simple procedure to eliminate this instability is to half the time step size until convergence in computed results is achieved. Generally,

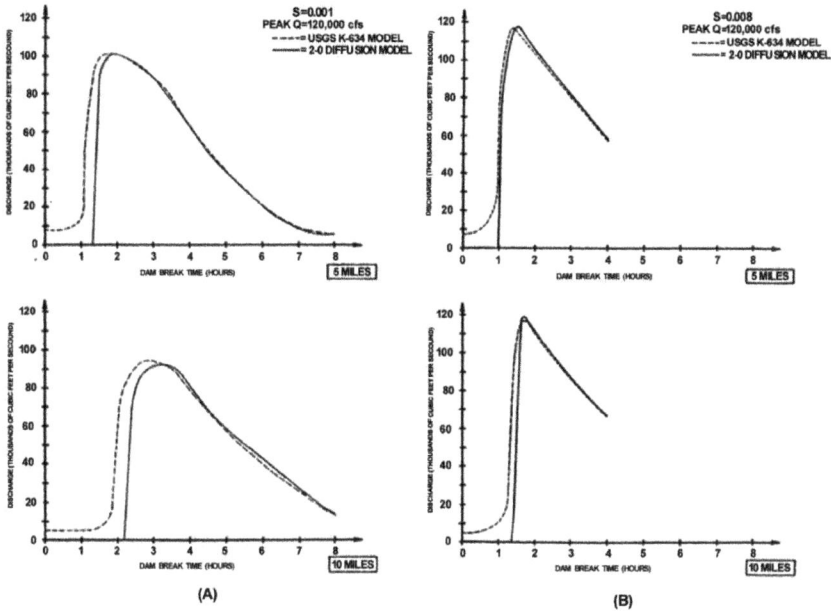

Figure 2.
Comparisons of outflow hydrographs at 5 and 10 miles downstream from the dam – break site (peak Q = 120,000 cfs) (A) S = 0.001 (B) S = 0.008.

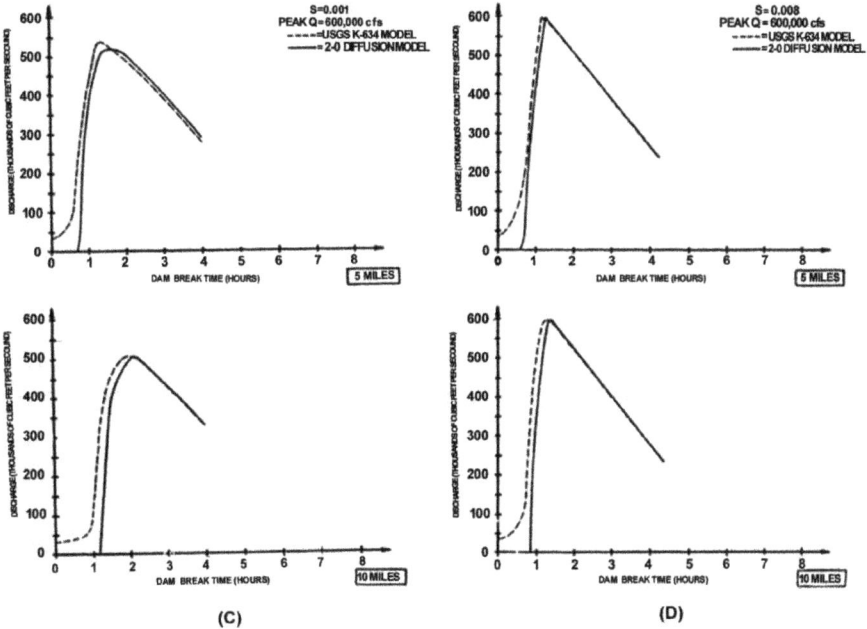

Figure 3.
Comparisons of outflow hydrographs at 5 and 10 miles downstream from the dam – break site (peak Q = 600,000 cfs) (C) S = 0.001 (D) S = 0.008.

such a time step adjustment may be directly included in the computer program for the dam-break model. For the cases considered in this section, the time step size of 7.2 s was found to be adequate when using the 1000–5000-feet grid sizes.

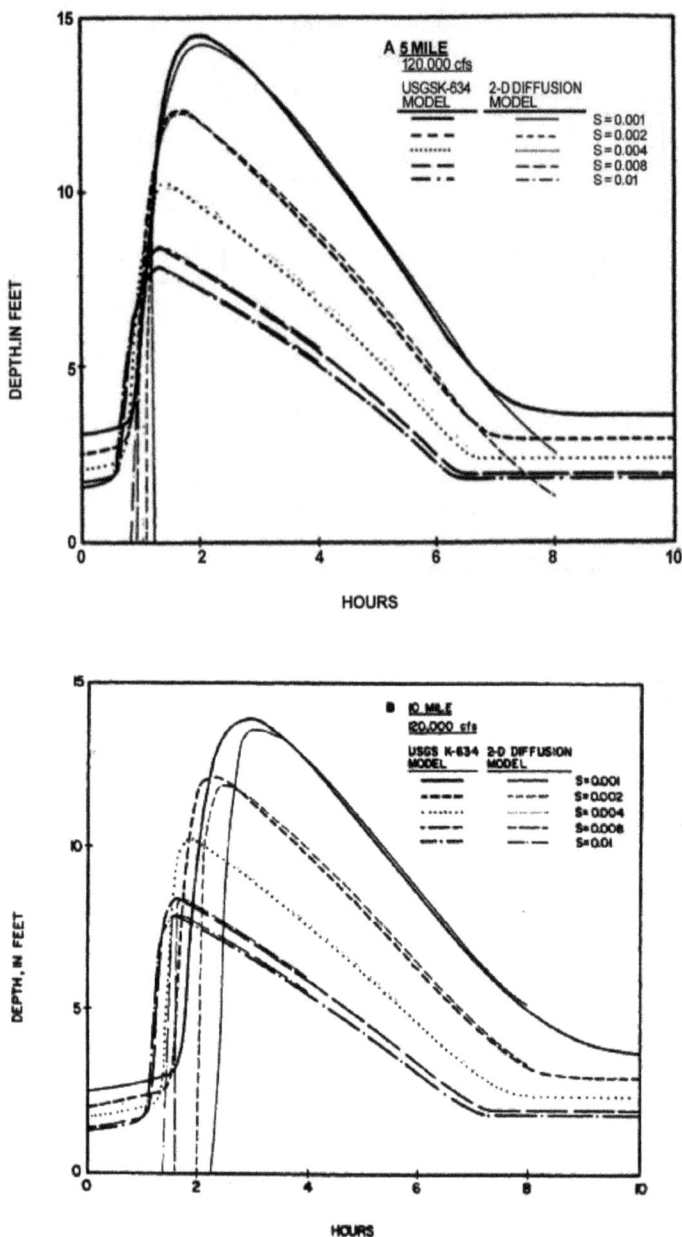

Figure 4.
Comparisons of depths of water at (A) 5 miles and (B) 10 miles downstream from the dam-break site (peak Q = 120,000 cfs).

2.3 Results

For the dam-break hydrographs considered and the range of channel slopes modeled, the simple diffusion dam-break model of Eq. (22) in Chapter 1 provides estimates of flood depths and outflow hydrographs which compare favorably to the results determined by the well-known K-634 one-dimensional dam-break model. Generally speaking, the difference between the two modeling approaches is found to be less than a 3% variation in predicted flood depths.

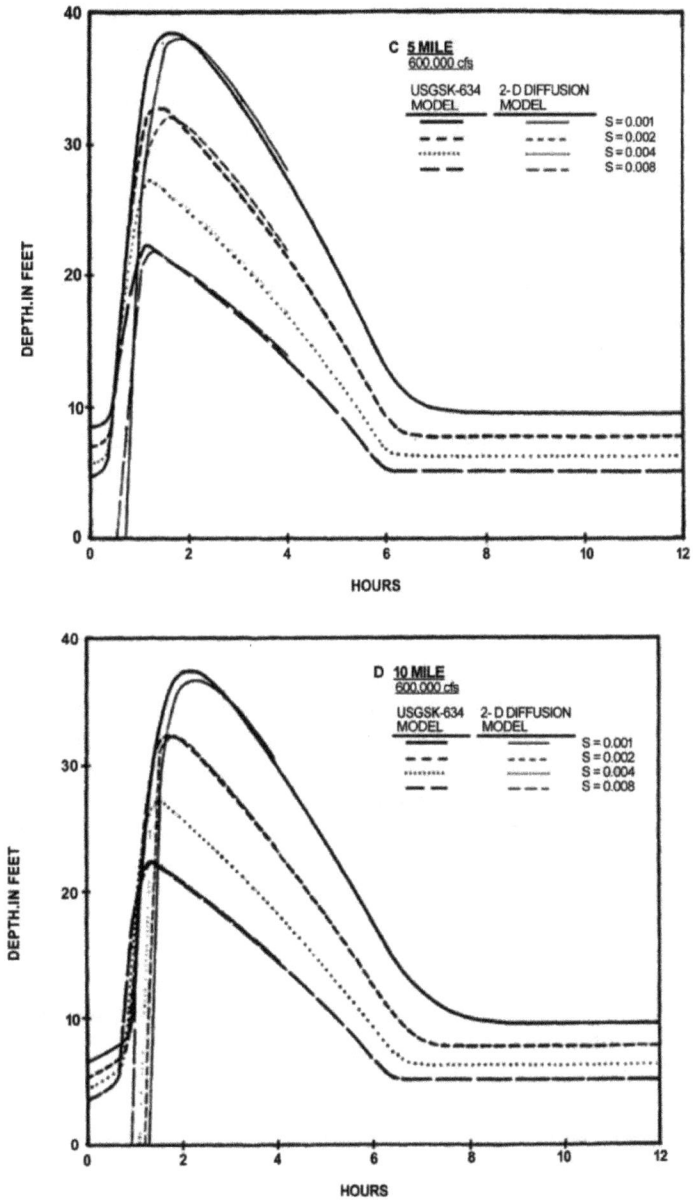

Figure 5.
Comparisons of depths of water at (C) 5 miles and (D) 10 miles downstream from the dam-break site (peak Q = 600,000 cfs).

The presented diffusion dam-break model is based upon a straightforward explicit time stepping method which allows the model to operate upon the nodal points without the need to use large matrix systems. Consequently, the model can be implemented on most currently available microcomputers. However, as compared to implicit solution methods, time steps for DHM use are extremely small. Thus, relatively short simulation times must be used.

The diffusion model of Eq. (22) in Chapter 1 can be directly extended to a two-dimensional model by adding the y-direction terms, which are computed in a similar fashion as the x-direction terms. The resulting two-dimensional diffusion model is texted by modeling the considered test problems in the x-direction, the y-direction, and along a 45-degree trajectory across a two-dimensional grid aligned with the x-y coordinate axis. Using a similar two-dimensional model, Xanthopoulos and Koutitas [9] conceptually verify the diffusion modeling technique by considering the evolution of a two-dimensional floodplain which propagates radially from the dam-break site.

From the above conclusions, the use of the diffusion approach (Chapter 1, Eq. 22), in a two-dimensional DHM may be justified due to the low variation in predicted flooding depths (one-dimensional) with the exclusion of the inertial terms. Generally speaking, a two-dimensional model would be employed when the expansion nature of flood flows is anticipated. Otherwise, one of the available one-dimensional models would suffice for the analysis.

3. Two-dimensional analysis

3.1 Introduction

In this section, a two-dimensional DHM is developed. The model is based on a diffusion approach where gravity, friction, and pressure forces are assumed to dominate the flow equations. Such an approach has been used earlier by Xanthopoulos and Koutitas [9] in the prediction of dam-break floodplains in Greece. In those studies, good results were also obtained by using the two-dimensional model for predicting one-dimensional flow quantities. In the preceding section, a one-dimensional diffusion model has been considered, and it has been concluded that for most velocity flow regimes (such as Froude number less than approximately 4), the diffusion model is a reasonable approximation of the full dynamic wave formulation.

An integrated finite difference grid model is developed which equates each cell-centered node to a function of the four neighboring cell nodal points. To demonstrate the predictive capacity of the floodplain model, a study of a hypothetical dam break of the Crowley Lake dam near the city of Bishop, California (**Figure 6**), is considered [18, 19].

The steepness and confinement of the channel right beneath the Crowley Lake dam results in a translation of outflow hydrograph in time. Therefore, the dam-break analysis is only conducted in the neighborhood near the city of Bishop, where the gradient of topography is mild.

3.2 K-634 modeling results and discussion

Using the K-634 model for computing the two-dimensional flow was attempted by means of the one-dimensional nodal spacing (**Figure 7**). Cross sections were obtained by field survey, and the elevation data were used to construct nodal point flow-width versus stage diagrams. A constant Manning's roughness coefficient of 0.04 was assumed for study purposes. The assumed dam failure reached a peak flow rate of 420,000 cfs within 1 h and returned to zero flow 9.67 h later. **Figure 8** depicts the K-634 floodplain limits. To model the flow breakout, a slight gradient was assumed for the topography perpendicular to the main channel. The motivation for such a lateral gradient is to limit the channel flood-way section in order to approximately conserve the one-dimensional momentum equations. Consequently, fictitious channel sides are included in the K-634 model study, which results in artificial confinement of the flows. Hence, a narrower floodplain is delineated in **Figure 8** where the flood flows are falsely retained within a hypothetical channel confine. An examination of the flood depths given in **Figure 9** indicates that at

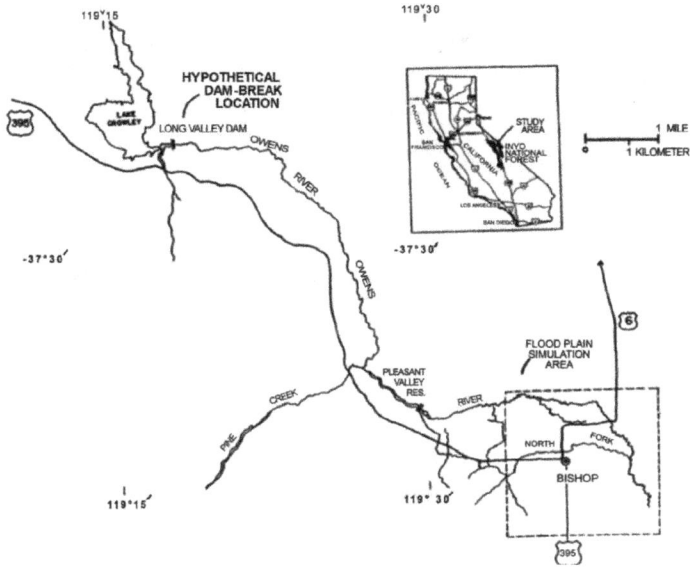

Figure 6.
Dam-break study location.

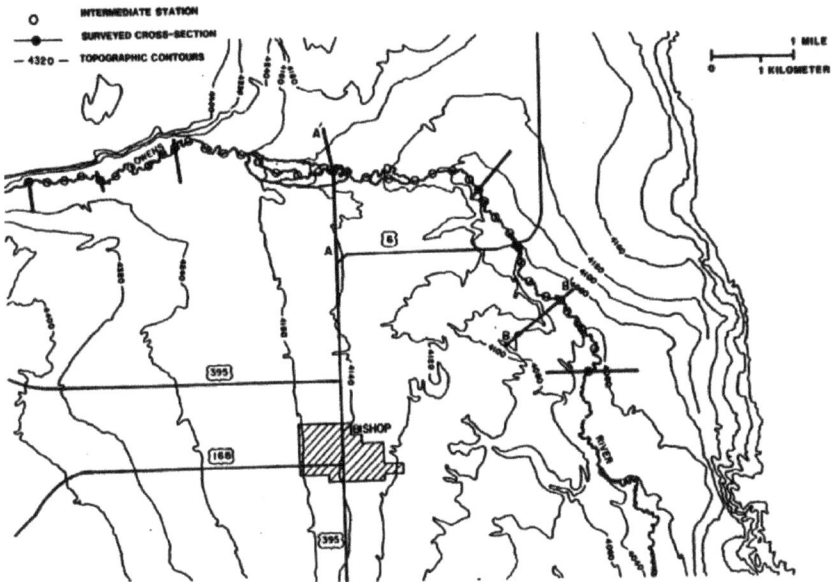

Figure 7.
Surveyed cross section locations on Owens River for use in K-634 model.

the widest floodplain expanse of **Figure 8**, the flood depth is about 6 feet, yet the floodplain is not delineated to expand southerly but is modeled to terminate based on the assumed gradient of the topography toward the channel. Such complications in accommodating an expanding floodplain when using a one-dimensional model are obviously avoided by using a two-dimensional approach.

The two-dimensional diffusion hydrodynamic model is now applied to the hypothetical dam-break problem using the grid discretization shown in **Figure 10**. The same

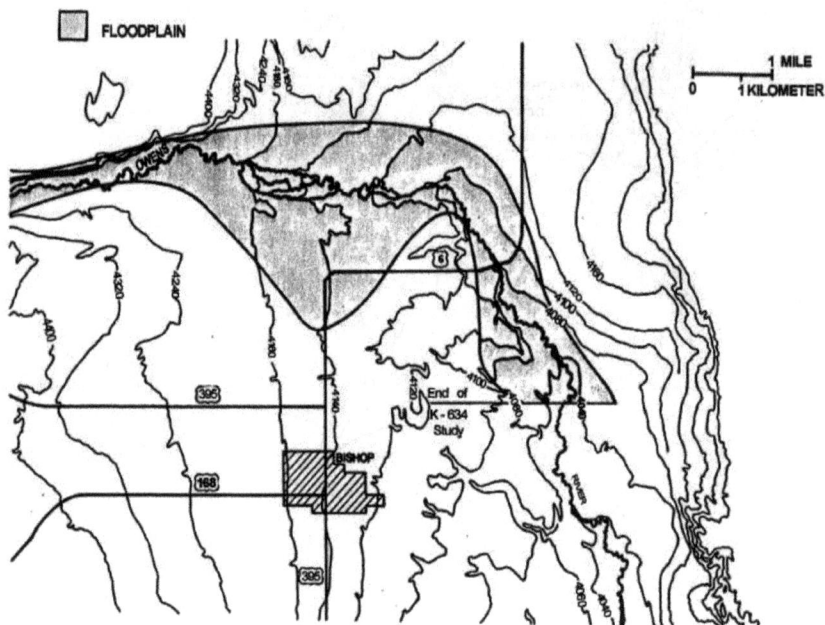

Figure 8.
Floodplain computed from K-634 model.

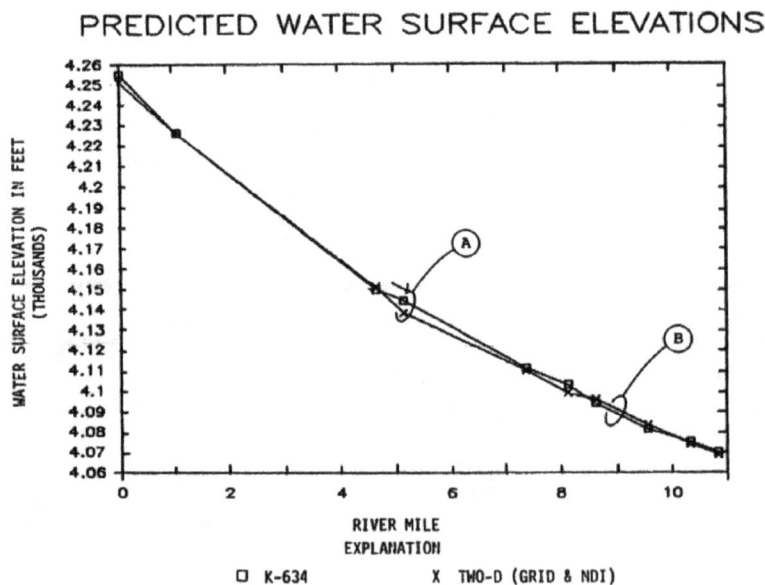

Figure 9.
Comparison of modeled water surface elevations (Points A and B in the figure are selected as example locations where a greater than an average difference between tested model predictions are observed).

inflow hydrograph used in K-634 model is also used for the diffusion hydrodynamic model. Again, Manning's roughness coefficient at 0.04 was used. The resulting floodplain is shown in **Figure 11** for the 1/4 square-mile grid model.

The two approaches are comparable except at cross sections shown as A-A and B-B in **Figure 7**. Cross section A-A corresponds to the predicted breakout of flows

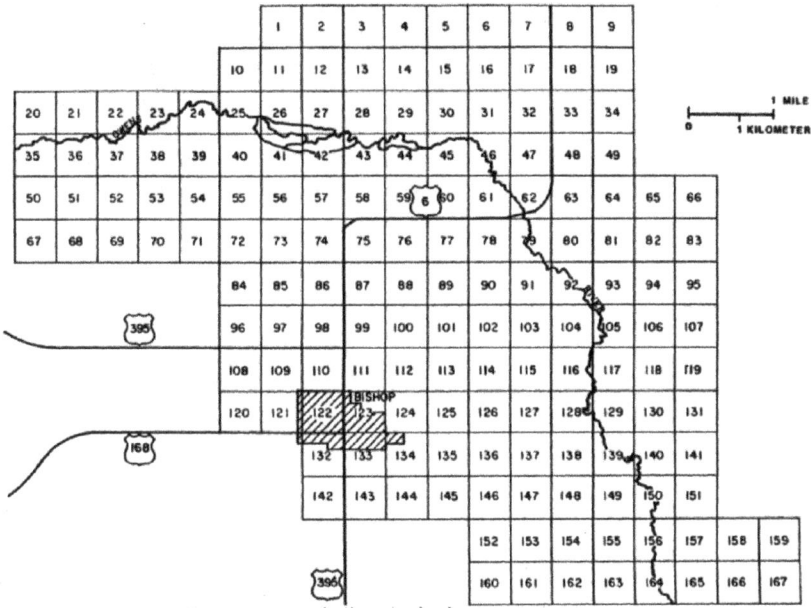

Figure 10.
Floodplain discretization for two-dimensional diffusion hydrodynamic model.

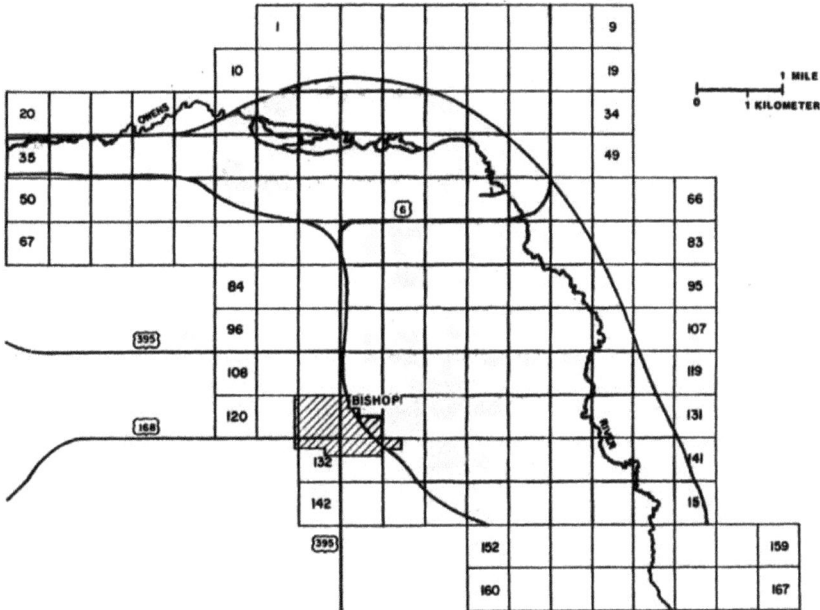

Figure 11.
Floodplain for two-dimensional diffusion hydrodynamic model.

away from the Owens River channel with flows traveling southerly toward the city of Bishop. As discussed previously, the K-634 predicted flood depth corresponds to a flow depth of 6 feet (above natural ground) which is actually unconfined by the channel. The natural topography will not support such a flood depth, and, consequently, there should be southerly breakout flows such as predicted by the two-dimensional model. With such breakout flows included, it is reasonable that the two-dimensional model would predict a lower flow depth at cross section A-A.

At cross section B-B, the K-634 model predicts a flood depth of approximately 2 feet less than the two-dimensional model. However, at this location, the K-634 modeling results are based on cross sections, which traverse a 90-degree bend. In this case K-634 model will overestimate the true channel storage, resulting in an underestimation of flow depths.

4. Conclusions

The contribution of inertial terms for one-dimensional flows resulting from a dam break was investigated by comparing the results of the DHM with the K-634 model, which includes inertial terms. The close agreement between the two models predicted results justifies the use of the DHM for these applications.

For two-dimensional flows, comparing the various model predicted flood depths and delineated plains, it is seen that the two-dimensional diffusion hydrodynamic model predicted more reasonable floodplain boundary, which is associated with broad, flat plains such as those found at the study site. The model approximates channel bends, channel expansions and contractions, flow breakouts, and the general area of inundation. Additionally, the diffusion hydrodynamic model approach allows for the inclusion of return flows (to the main channel), which were the result of upstream channel breakout, and other two-dimensional flow effects, without the need for special modeling accommodations that would be necessary with using a one-dimensional model.

Author details

Theodore V. Hromadka II [1*] and Chung-Cheng Yen[2]

1 Department of Mathematical Sciences, United States Military Academy, West Point, NY, USA

2 Tetra Tech, Irvine, CA, USA

*Address all correspondence to: ted@phdphdphd.com

IntechOpen

References

[1] Land LF. Mathematical simulations of the Toccoa Falls, Georgia, Dam-Break Flood. Water Resources Bulletin. 1980;**16**(6):1041-1048

[2] Land LF. Evaluation of selected dam-break flood-wave models by using field data. U.S.G.S. Water Resources Investigations. 1980;**80**(4):54

[3] Ponce VM, Tsivoglou AJ. Modeling gradual dam breaches. ASCE Journal of Hydrauic Division. 1981;**107**(7):829-838

[4] Rajar R. Mathematical simulation of dam-break flow. ASCE Journal of Hydrauic Division. 1978;**104**(7):1011-1026

[5] Sakkas JG, Strelkoff T. Dam-break flood in a prismatic dry channel. ASCE Journal of Hydrauic Division. 1973;**99**(12):2195-2216

[6] Chen C. Laboratory verification of a dam-break flood model. ASCE Journal of Hydrauic Division. 1980;**106**(4):535-556

[7] Chen C, Armbruster JT. Dam-break wave model: Formulation and verification. ASCE Journal of Hydrauic Division. 1980;**106**(5):747-767

[8] Katopodes N, Strelkoff T. Computing two-dimensional dam-break flood waves. ASCE Journal of Hydrauic Division. 1978;**104**(9):1269-1288

[9] Xanthopoulos T, Koutitas C. Numerical simulation of a two-dimensional flood wave propagation due to dam failure. ASCE Journal of Hydraulic Research. 1976;**14**(4):321-331

[10] Akan AO, Yen BC. Diffusion-wave flood routing in channel networks. ASCE Journal of Hydrauic Division. 1981;**107**(6):719-732

[11] Hunt T. Asymptotic solution for dam-break problem. ASCE Journal of Hydrauic Division. 1982;**108**(1):115-126

[12] Ponce VM, Li RM, Simons DB. Applicability of kinematic and diffusion models, in verification of mathematical and physical models in hydraulic engineering. In: American Society of Civil Engineers, Hydraulic Division, Special Conference College Park, Maryland: University of Maryland; 1978. pp. 605-613

[13] Ponce VM. Nature of wave attenuation in Open Channel flow. ASCE Journal of Hydrauic Division. 1982;**108**(2):257-262

[14] Miller WA, Cunge JA. Simplified Equations of Unsteady Flow. Chapter 5 of Unsteady Flow in Open Channels. Vol. 1. Fort Collins, Colorado: Water Resources Publications; 1975. pp. 183-257

[15] Morris EM, Woolhiser DA. Unsteady one-dimensional flow over a plane: Partial equilibrium and recession hydrographs. Water Resources Research, AGU. 1980;**16**(2):356-360

[16] Henderson FM. Flood waves in prismatic channels. ASCE Journal of Hydrauic Division. 1963;**89**(4):39-67

[17] Basco DR. Introduction to numerical method—Part I and II. Verification of mathematical and physical models in hydraulic engineering. In: American Society of Civil Engineers, Hydraulic Division, Special Conference College Park, Maryland: University of Maryland; 1978. pp. 280-302

[18] Hromadka II TV, Lai C. Solving the two-dimensional diffusion flow model. In: Proceedings of ASCE Hydraulics Division Specialty Conference, Orlando, Florida. 1985

[19] Hromadka II TV, Nestlinger AB. Using a two-dimensional diffusional dam-break model in engineering planning. In: Proceedings of ASCE Workshop on Urban Hydrology and Stormwater Management, Los Angeles County Flood Control District Office, Los Angeles, California. 1985

Chapter 3

Program Description of the Diffusion Hydrodynamic Model

Theodore V. Hromadka II and Chung-Cheng Yen

Abstract

The numerical algorithm, with a focus on the interface element that is used in the diffusion hydrodynamic model, is presented in this chapter. The source program was written in FORTRAN language, and it can be downloaded from this book companion website. The channel, flood plain, and the interface flow details are discussed.

Keywords: flood plain, overflow, control volume, water elevation, channel interface

1. Introduction

A computer program for the two-dimensional diffusion hydrodynamic model which is based on the diffusion form of the St. Venant equations [1–5] where gravity, friction, and pressure forces are assumed to dominate the flow equation will be discussed in this section.

The DHM consists of a 1-D channel and 2-D flood plain models, and an interface sub-model. The one-dimensional channel element utilizes the following assumptions:

- Infinite vertical extensions on channel walls (**Figure 1**)

- Wetted perimeter is calculated as shown in **Figure 1(a)**

- Volumes due to channel skew are ignored (**Figure 1(b)**)

- All overflow water is assigned to one grid element (**Figure 2**)

The interface model calculates the excess amount of water either from the channel element or from the flood plain element. This excess water is redistributed to the flood plain element or the channel element according to the water surface elevation.

This FORTRAN program has the capabilities to simulate both one-and two-dimensional surface flow problems, such as the one-dimensional open channel flow and two-dimensional dam-break problems illustrated in the preceding pages. Engineering applications of the program will be presented in the next chapter.

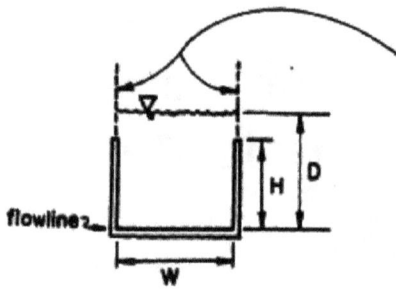

vertical extensions

Wetted perimeter (P)

(a)

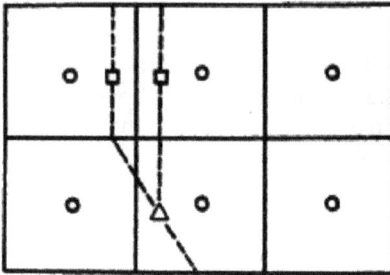

Legend

o - grid node

□ - channel node

Δ - channel junction

(b)

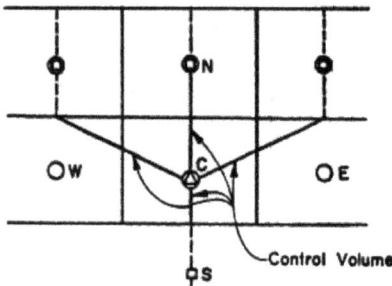

Assumptions

- Ignore volume differences due to channel skew

-All overflow assigned to one grid element (see interface model)

(c)

Figure 1.
One-dimensional channel flow element characteristics: (a) element geometrics, (b) element associations to grid elements, and (c) channel element connections.

2. Interface model

2.1 Introduction

The interface model modifies the water surface elevations of flood plain grids and channel elements at specified time intervals (update intervals). There are three cases of interface situations: (1) channel overflow, (2) grid overflow, and (3) flooding of channel and grid elements.

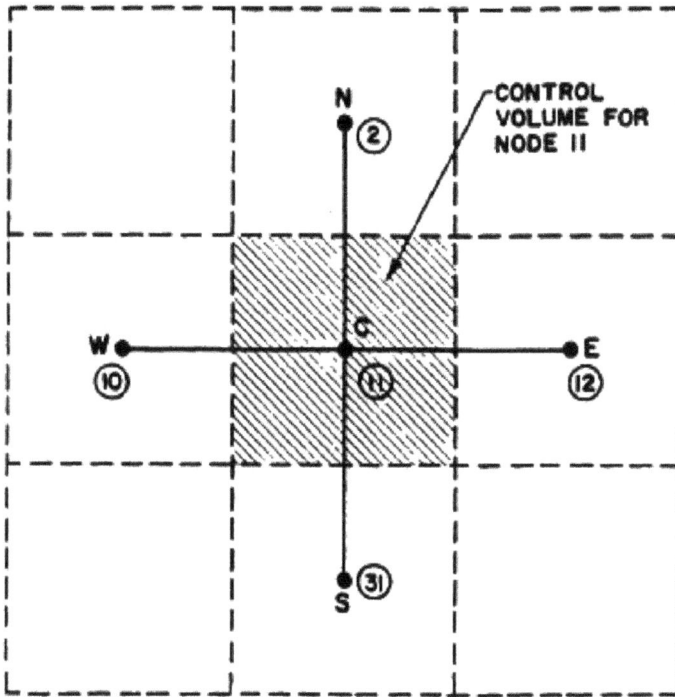

Figure 2.
Grid element nodal molecule.

2.2 Channel overflow

When the channel is overflowing; the excess water is temporarily stored in the vertically extended space (**Figure 3(b)**). Actually, it is the volume per unit length. This excess water is the product of the depth of water, the width of the channel, and the length of the channel and is subsequently uniformly distributed over the grid elements. In other words, the new grid water surface elevation is equal to the old water surface elevation plus a depth of hw/L, and the channel water surface elevation now matches the parent grid water surface elevation.

2.3 Grid overflow

When the water surface elevation of the grid element is greater than a specified surface detention (**Figure 3(a)**), the excess water drains into the channel element, and the new water surface elevation is changed according to the following two conditions (**Figure 3(c)**), (a) if $v > v'$, where v denotes the excess volume of water per unit length and v' denotes the available volume per unit length, the new water surface of the grid element is $A^{NEW} = A^{OLD} - (v - v')/L$, and the new water surface elevation of the channel element is also equal to A^{NEW}, and (b) if $v \leq v'$, the new water surface elevation of the grid element is $A^{NEW} = A^{OLD} - h$ and the new water surface elevation of the channel element is $B^{NEW} = B^{OLD} + v/w$.

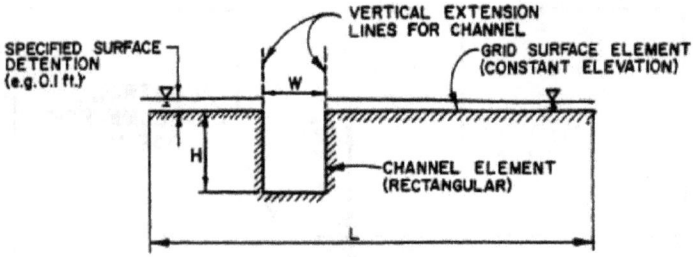

(a)

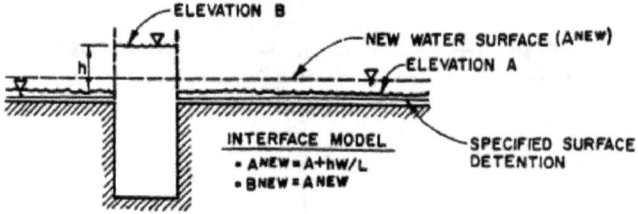

(b)

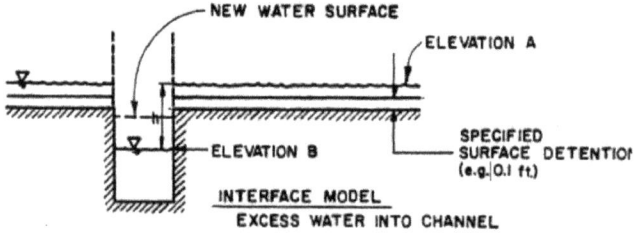

(c)

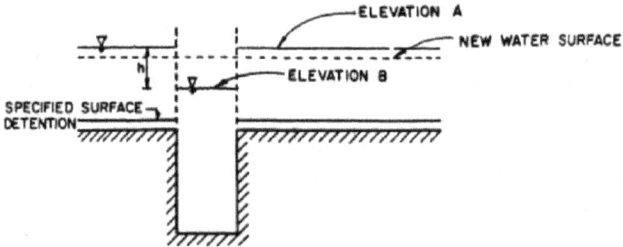

(d)

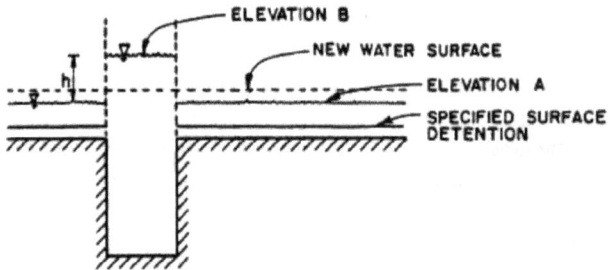

(e)

Figure 3.
Diffusion hydrodynamic interface model: (a) model interface geometries, (b) channel overflow interface model, (c) grid overflow interface model, (d) grid-channel flooding interface model, and (e) channel-grid flooding interface model.

2.4 Flooding of channel and grid

When flooding occurs, the water surface elevations of the grid and channel elements are both greater than the specified surface detention elevation. Two cases have to be considered as follows:

1. If A > B (**Figure 3(d)**), the new water surface elevation of the grid element is
$A^{NEW} = B^{OLD} + \dfrac{h(L-w)}{L}$, and the new water surface elevation of the channel
element is equal to A^{NEW}.

2. If A < B (**Figure 3(e)**), the new water surface elevation of the grid element is
$A^{NEW} = A^{OLD} + \dfrac{h.w}{L}$, and the new water surface elevation of the channel element
is equal to A^{NEW}.

3. Conclusions

The characteristic features of the diffusion hydrodynamic model are detailed with a focus on its ability to model the interface. The interface component in the model can modify the water elevation in the grids along the flood plain and channel to account for different overflow scenarios, which are also illustrated.

Author details

Theodore V. Hromadka II [1*] and Chung-Cheng Yen[2]

1 Department of Mathematical Sciences, United States Military Academy, West Point, NY, USA

2 Tetra Tech, Irvine, CA, USA

*Address all correspondence to: ted@phdphdphd.com

IntechOpen

References

[1] Abbott MB, Bathurst JC, Cunge JA, O'Connell PE, Rasmussen J. An introduction to the European hydrological system—Système Hydrologique Européen, "SHE", 1: History and philosophy of a physically based distributed modeling system. Journal of Hydrology. 1986;**87**(1-2):45-59

[2] Amein M, Fang CS. Implicit flood routing in natural channels. Journal of Hydraulics Division, Proceedings of ASCE. 1970;**96**(12):2481-2500

[3] Balloffet A, Scheffler ML. Numerical analysis of the Teton dam failure flood. Journal of Hydraulic Research. 1982;**20**(4):317-328

[4] Borah DK, Prasad SN, Alonso CV. Kinematic wave routing incorporating shock fitting. Water Resources Research. 1980;**16**(3):529-541

[5] Li RM, Simons DB, Stevens MA. Nonlinear kinematic wave approximation for water routing. Water Resources Research. 1975;**11**(2):245-252

Chapter 4

Applications of Diffusion Hydrodynamic Model

Theodore V. Hromadka II and Chung-Cheng Yen

Abstract

The diffusion hydrodynamic model is applied for seven engineering applications that are commonly encountered in real-world applications. Of the seven applications, six relate to two-dimensional flows. The results are compared to other model results, where available. The results underscore the reliability of the DHM along with its limitation for modeling rapidly varying flows.

Keywords: hydraulic jump, hydrograph, two-dimensional flows, estuary, channel flood plain interface

1. One-dimensional model application

1.1 Application 1: steady flow in an open channel

Because the DHM is anticipated for use in modeling watershed phenomena, it is important that the channel models represent known flow characteristics. Unsteady flow is examined in the previous section. For steady flow, a steady-state, gradually varied flow problem is simulated by the 2-D diffusion model. **Figure 1** depicts both the water levels form the 2-D diffusion model and from the gradually varied flow equation. For an 8000 cfs constant inflow rate, the water surface profiles from both the 2-D diffusion model and the gradually varied flow equation match quite well. The discrepancies of these profiles occur at the breakpoints where the upstream channel slope and downstream channel slope change. At the first break point where the upstream channel slope is equal to 0.001 and the downstream channel slope is equal to 0.005, the water surface level is assumed to be equal to the critical depth. However, Henderson [1] notes that brink flow is typically less than the critical depth (Dc). The DHM water surface closely matches the 0.72 Dc brink depth.

It is clear to see that the DHM cannot simulate the hydraulic jump but rather smooth out the usually assumed "shock front." However, when considering unsteady flow, the DHM may be a reasonable approach for approximating the jump profile. For a higher inflow rate, 20,000 cfs, the surface water levels differ in the most upstream reach. Again, this is due to the downstream control, critical depth, of the gradually varied flow equation.

IntechOpen

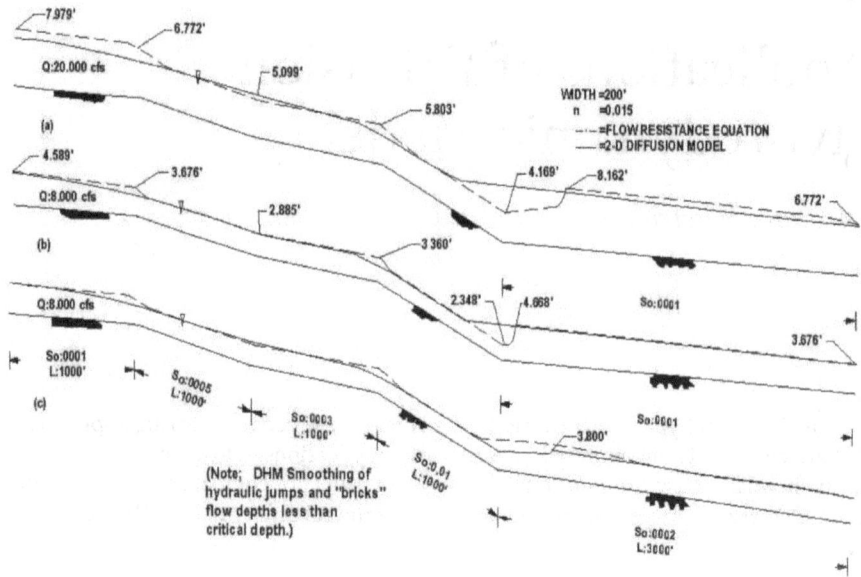

Figure 1.
Gradually varied flow profiles. (a) Q = 20,000 cfs, downstream slope = 0.0001 (b) Q = 8,000 cfs, downstream slope = 0.0001 (c) Q = 8,000 cfs, downstream slope = 0.0002.

2. Two-dimensional model applications

2.1 Application 2: rainfall-runoff model

The DHM can be used to develop a runoff hydrograph given the time distribution of effective rainfall. To demonstrate the DHM runoff hydrograph generation
[2], the DHM is used to develop a synthetic S-graph for a watershed where the
overland flow is the dominating flow effect.

To develop the S-graph, a uniform effective rainfall is assumed to uniformly
occur over the watershed. For each time step (5 s), an incremental volume of water
is added directly to each grid element based on the assumed constant rainfall
intensity, resulting in an equivalent increase in the nodal point depth of water.
Runoff flows to the point of concentration according to the two-dimensional
diffusion hydrodynamics model.

The 10 square mile Cucamonga Creek Watershed (California) is shown,
discretized by 1000-foot grid elements, in **Figure 2**. A design storm (**Figure 3**) was
applied to the watershed, and the resulting runoff hydrographs are depicted in
Figure 4 for DHM and synthetic unit hydrograph method. From **Figure 4**, the
diffusion model generates runoff quantities which are in good agreement with the
values computed using a synthetic unit hydrograph method derived from stream
gage data.

Next, the DHM is applied to three hypothetical dam failures in Orange
County, California (see **Figure 5**). In this application, the ability of DHM to
predict flow characteristics in domains where flood flow patterns are affected by
railroad, the bridge under crossings, and other man-made obstacles to flow is
illustrated.

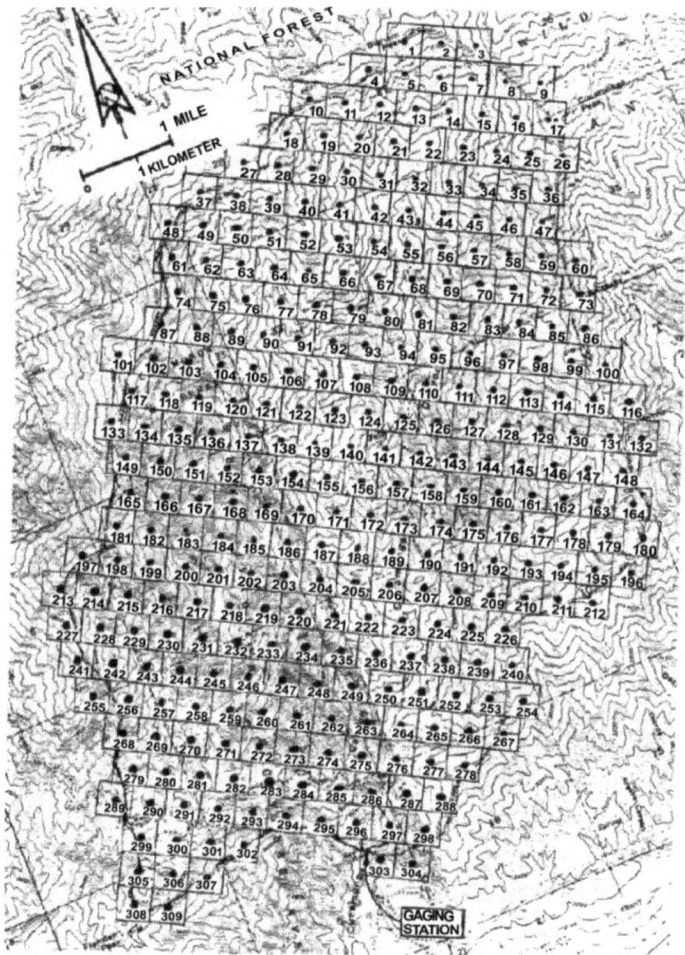

Figure 2.
Cucamonga Creek discretization.

Major assumptions used in these assumptions are as follows:

1. In each grid, an area-averaged ground elevation was estimated based on the topographic map, and a Manning's roughness coefficient was used for each application.

2. All storm drain systems provide negligible draw off of the dam-break flows. This assumption accommodates a design storm in progress during the dam failure. This assumption also implies that stormwater runoff provides a negligible increase to the dam-break flow hydrograph.

3. All canyon damming effects due to culvert crossings provide negligible attenuation of dam-break flows. This assumption is appropriate due to the concurrent design storm assumption and due to sediment deposition from the transport of the reservoir earthen dam materials.

4. The reservoir failure yields an outflow hydrograph, as depicted in **Figure 6**.

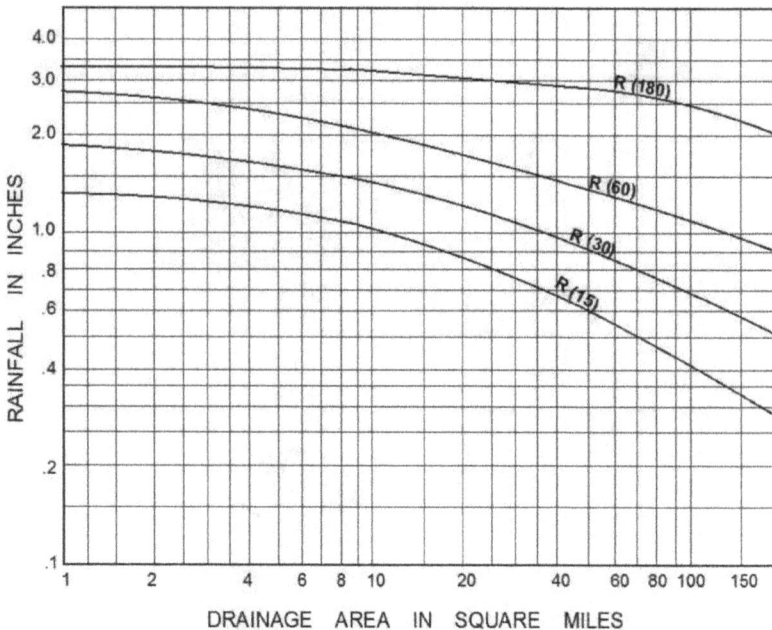

DRAINAGE AREA IN SQUARE MILES

HYETOGRAPH COMPUTATION

UNIT PERIOD	AMOUNT	
1	.07	(R(180)-R(60))
2	.05	(R(180)-R(60))
3	.11	(R(180)-R(60))
4	.05	(R(180)-R(60))
5	.20	(R(180)-R(60))
6	.22	(R(180)-R(60))
7	.14	(R(180)-R(60))
8	.16	(R(180)-R(60))
9	.48	(R(60)-R(30))
10	.52	(R(60)-R(30))
11	1.00	(R(15))
12	1.00	(R(30)-R(15))

Note:
R(15) - Rainfall (inches) in 15 minute curation

LOCAL PROJECT STORM
DEPTH AREA DURATION CURVES

Figure 3.
Design storm for Cucamonga Creek.

2.2 Application 3: small dam-break floodplain analysis

A study of a hypothetical failure of the Orange Country Reservoir northeast of the city of Brea, California (**Figure 7**), was conducted by Hromadka and Lai [3]. Using the USGS topographic quadrangle map (photo-revised, 1981), a 500-foot grid discretization was prepared (**Figure 8**), and nodal-area ground elevations were estimated based on the map. A Manning's roughness coefficient of n = 0.040 was used throughout the study, except in canyon reaches and grassy plains, where n was selected as 0.030 and 0.050, respectively. In this study, the resulting flood plain and the comparison of the model-simulated flood plain to a previous study by the Metropolitan Water District of Southern California [4] are shown in **Figure 9**. The main difference in the estimated flood plains is due to the dynamic nature of the

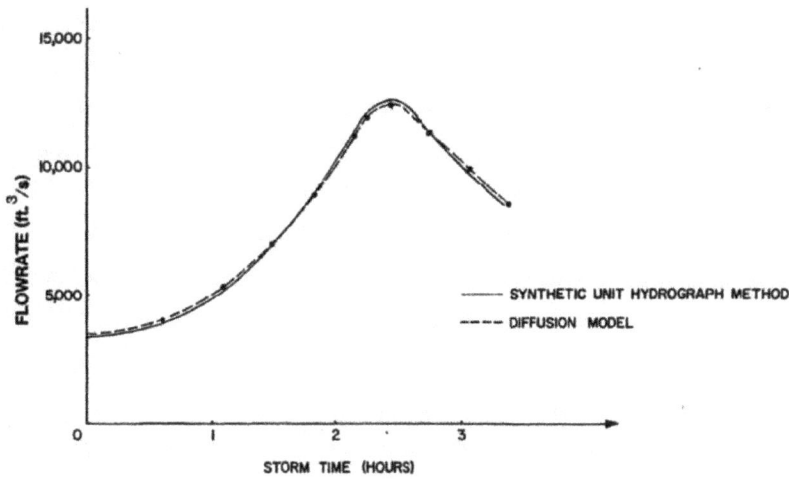

Figure 4.
Simulated runoff hydrographs for Cucamonga Creek.

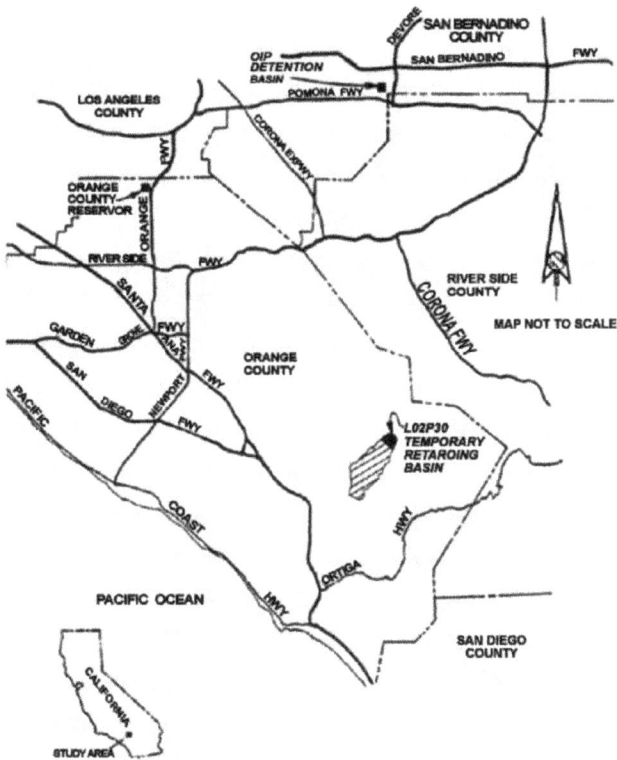

Figure 5.
Vicinity map for dam-break analyses.

DHM, which accounts for the storage effects resulting from flooding, and the attenuation of a flood wave because of 2-D routing effects. From this study, the estimated flood plain is judged to be reasonable.

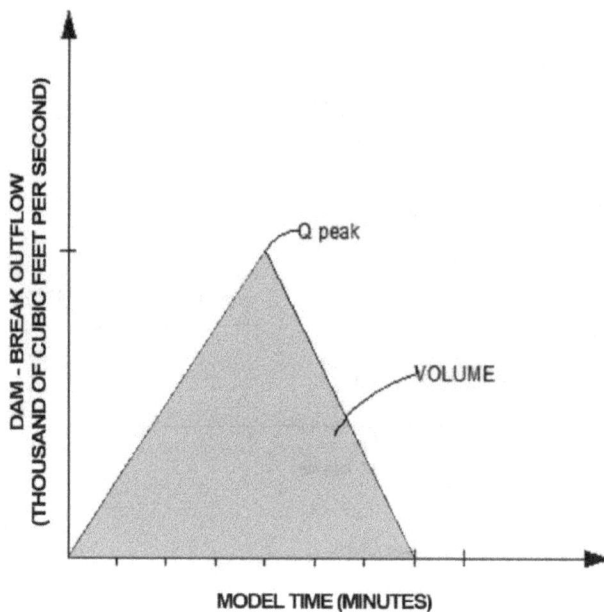

Figure 6.
Study dam-break outflow hydrograph for Orange County Reservoir.

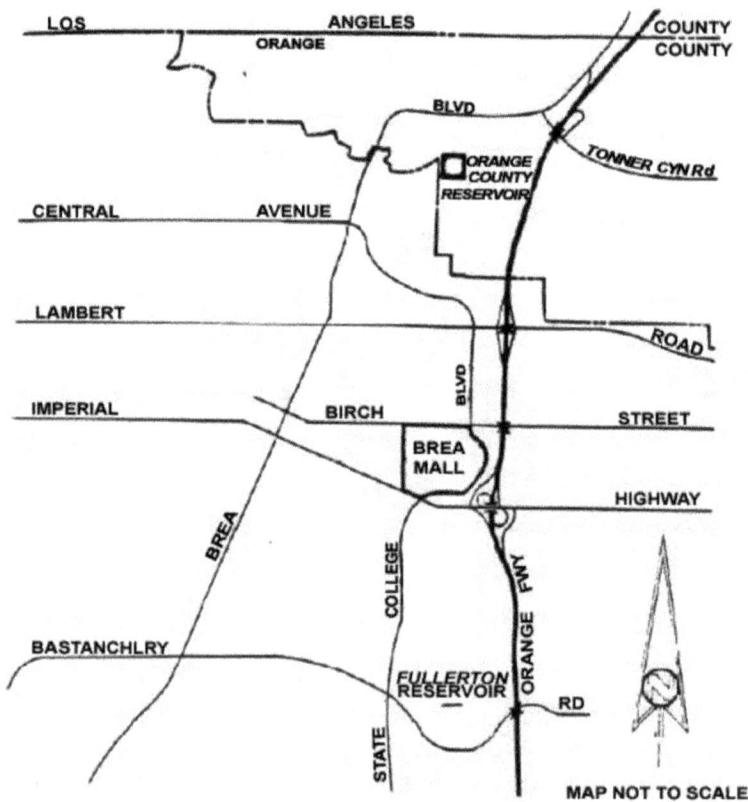

Figure 7.
Location map for the Orange County Reservoir dam-break problem.

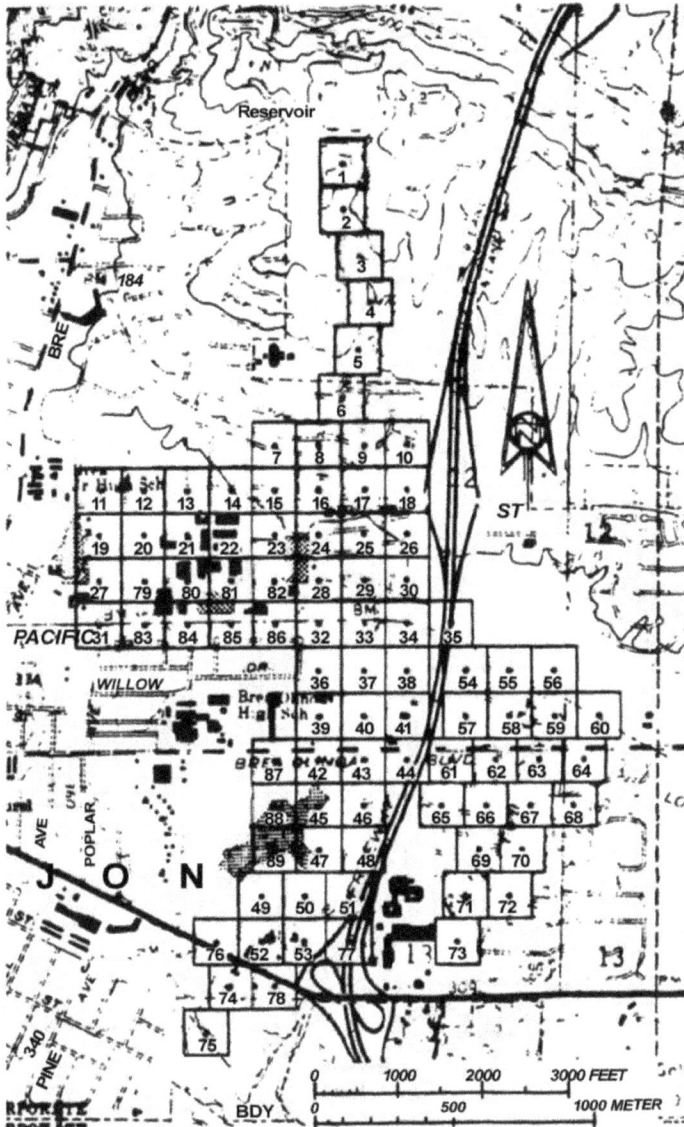

Figure 8.
Domain discretization for Orange County Reservoir.

2.3 Application 4: small-scale flows onto a flat plain

A common civil engineering problem is the use of temporary detention basins to offset the effects of urbanization on watershed runoff. A problem, however, is the analysis of the basin failure, especially, when the flood flows enter a wide expanse of land surface with several small channels. This application is to present study conclusions in estimating the flood plain, which may result from a hypothetical dam failure of the LO2P3O temporary retarding basin. The results of this study are to be used to estimate the potential impacts of the area if the retention basin berm were to fail.

The study site includes the area south of Plano Trabuco, Phase I. It is bounded on the north of LO2P3O Retarding Basin Berm, on the east and

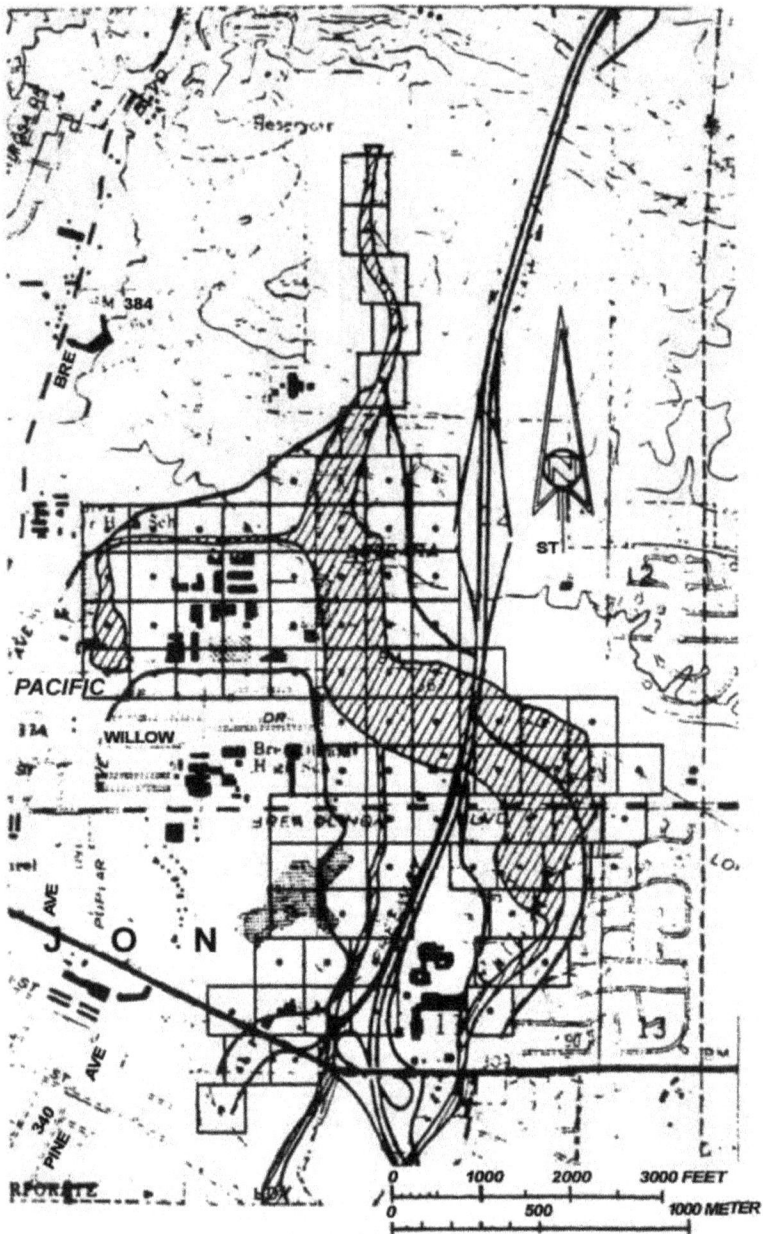

Figure 9.
Comparison of flood plain results for Orange County Reservoir.

south of Portola Parkway, and on the west by the Arroyo Trabuco bluffs (see **Figure 10**).

Using a 1″ = 300′ topographic map, a 200-foot grid control volume discretization was constructed, as shown in **Figure 11**. In each grid, an area-averaged ground elevation was estimated based on the topographic map. A Manning's roughness coefficient of n = 0.030 was used throughout the study.

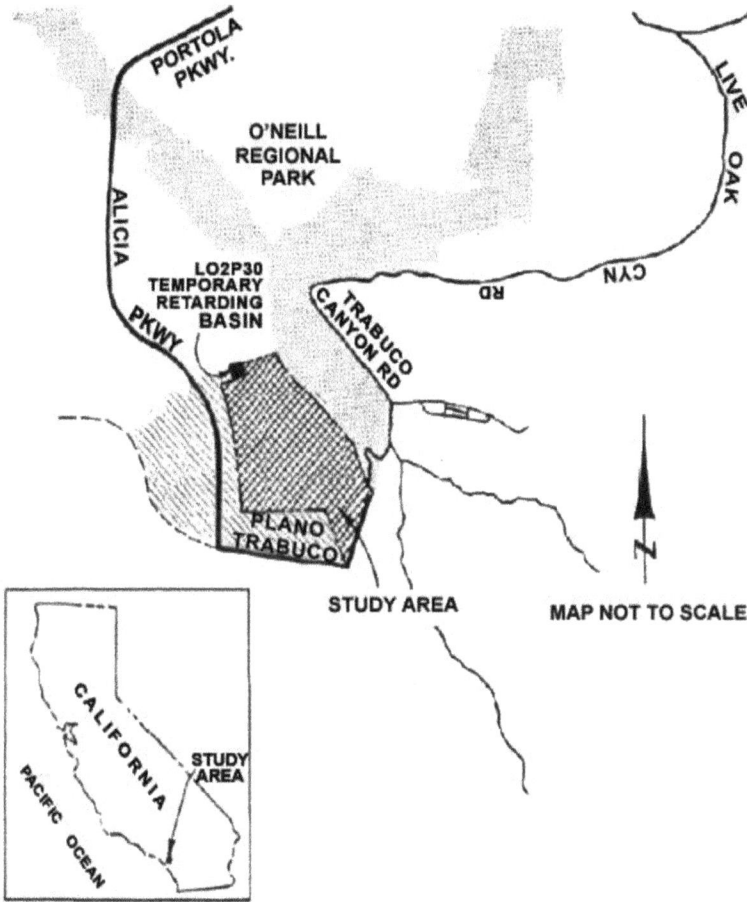

Figure 10.
Location map for L02P30 temporary retarding basin.

The profile of Portola Parkway varies approximately 2 ft above and below the adjacent land. Consequently, minor ponding may occur where Portola Parkway is high and sheet flow across Portola Parkway will occur at low points. It should be noted that depths along Portola Parkway are less than 1 foot (**Figure 11**). **Figure 12** shows lines of arrival times for the basin study. It is concluded that Portola Parkway is essentially unaffected by a hypothetical failure of the LO2P3O temporary retarding basin.

2.4 Application 5: two-dimensional flood flows around a large obstruction

In another temporary detention basin site, flood flows (from a dam-break) would pond upstream of a landfill site, and then split, when waters are deep enough, to flow on either side of the landfill. An additional complication is a railroad berm located downstream of the landfill, which forms a channel for flood flows. The study site (see **Figure 13**) is bounded on the north by a temporary berm approximately 300 ft north of the Union Pacific Railroad, bounded on the east by Milliken Avenue, bounded on the south by the Union Pacific Railroad, and bounded on the west by Haven Avenue.

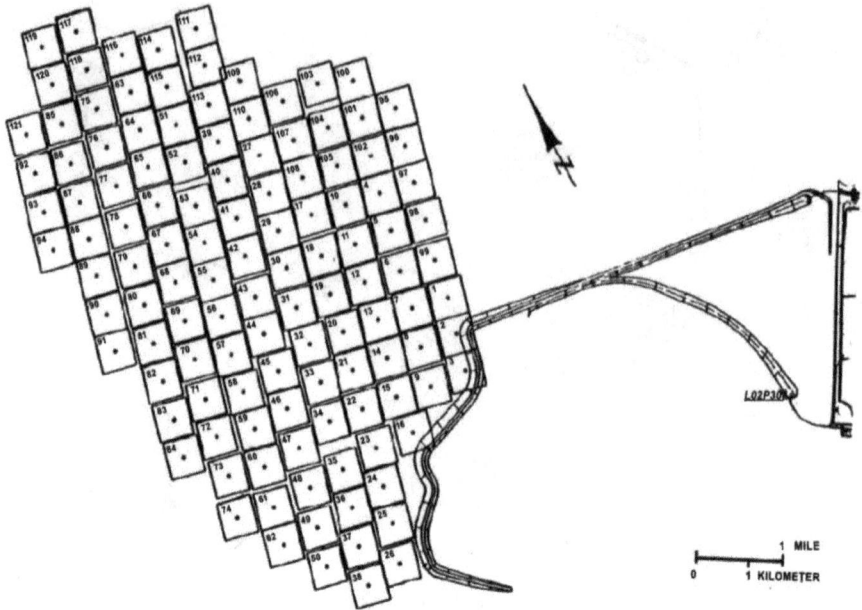

Figure 11.
Domain discretization of L02P30 temporary retarding basin.

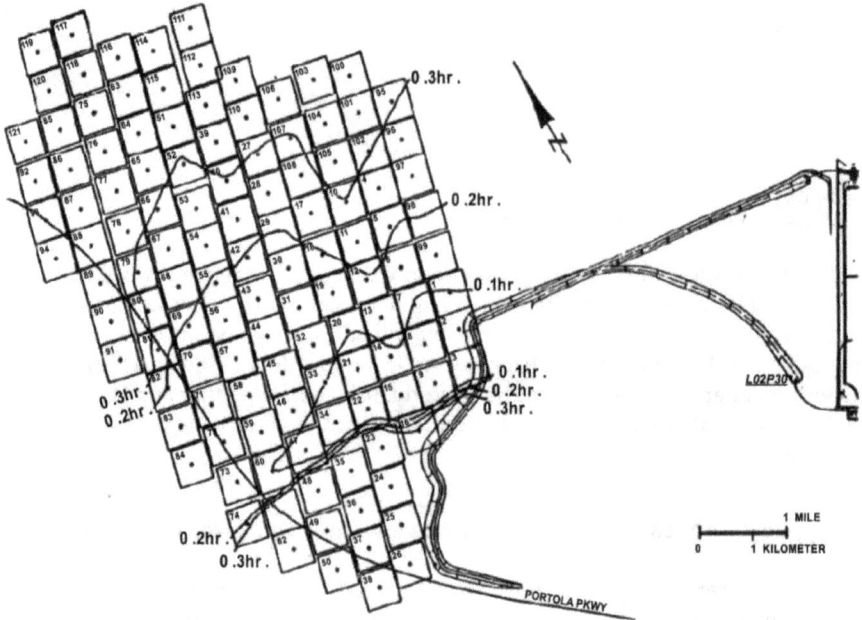

Figure 12.
Time of maximum flooding depth (80.5 acre—Feet basin test) for L02P30 temporary retarding basin.

A 200-foot grid control volume discretization was constructed as depicted in **Figure 14**. In each grid, an area-averaged ground elevation was estimated based on the topographic map. A Manning's roughness coefficient of n = 0.030 was used throughout the study.

Figure 13.
Location map for Ontario industrial partners' temporary detention basin.

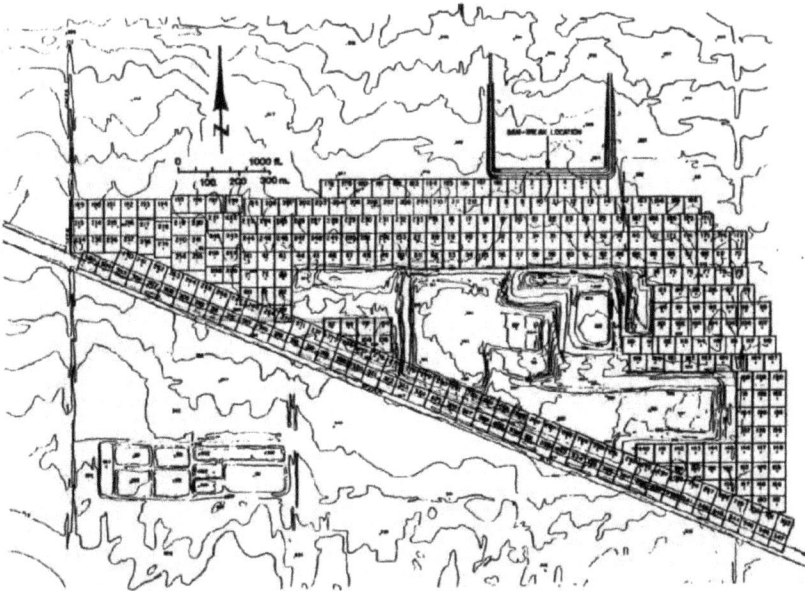

Figure 14.
Domain discretization for Ontario industrial partners' detention basin.

From **Figure 15**, it is seen that flood plain spreads out laterally and flows around the landfill. The flow ponds up around the landfill; along the north side of the landfill, the water ponds as high as 9.2 ft, and along the east and west sides of the

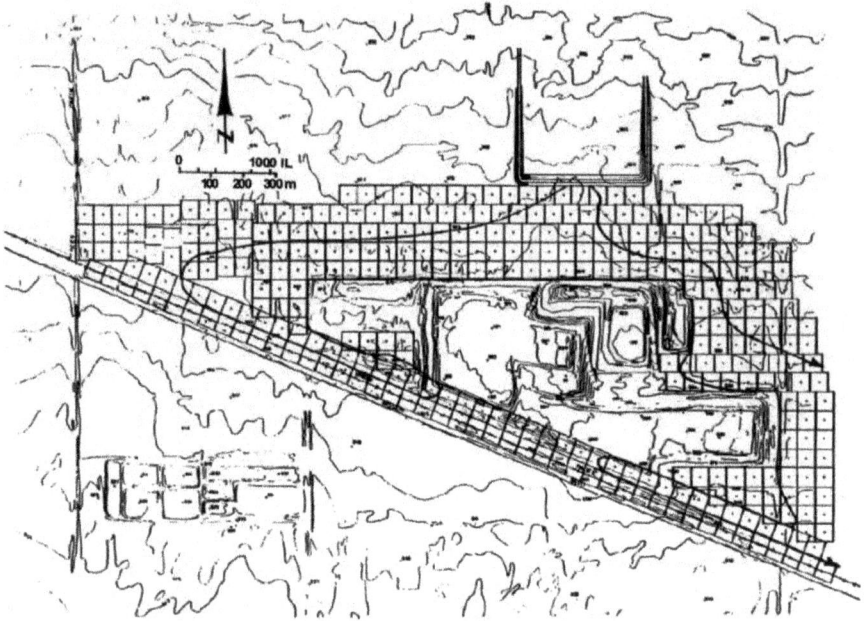

Figure 15.
Flood plain for Ontario industrial partners' detention basin.

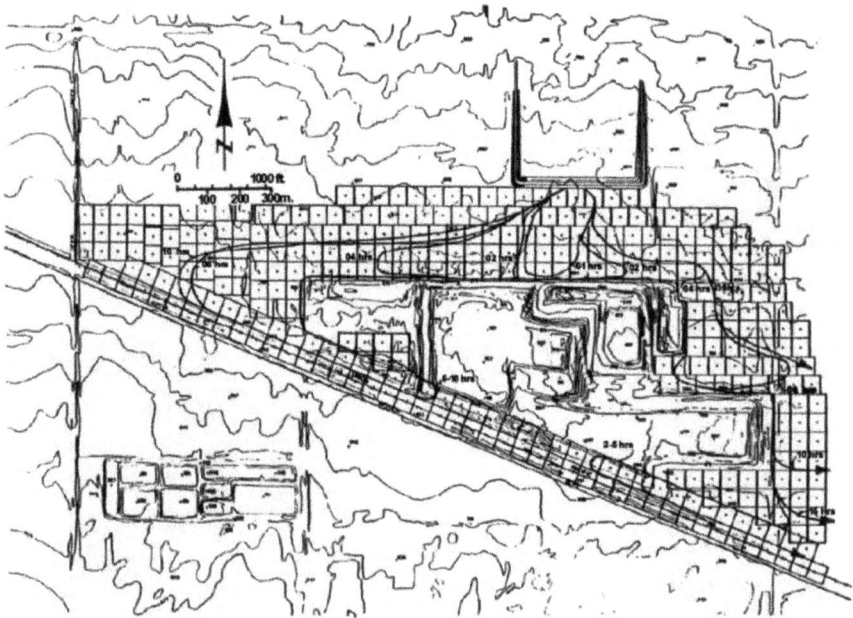

Figure 16.
Time (h) of maximum flooding depth for Ontario industrial partners detention basin.

landfill, the water ponds up to 5.1 ft high. As the flow travels south, it ponds up to a depth of 4.8 ft against the railroad near Milliken Avenue. Because the water spreads laterally, Milliken Avenue runs the risk of becoming flooded; however, the water only ponds to 0.6 ft along the street. A more in-depth study is needed to see if the water would remain in the gutter or flood Milliken Avenue.

By observing the arrival times of the flood plain in **Figure 16**, it is seen that the flood plain changes very little on the west side of the landfill once it reaches the railroad (0.6 h after the dam-break). But on the east side of the landfill, it takes 2.0 h to reach the railroad.

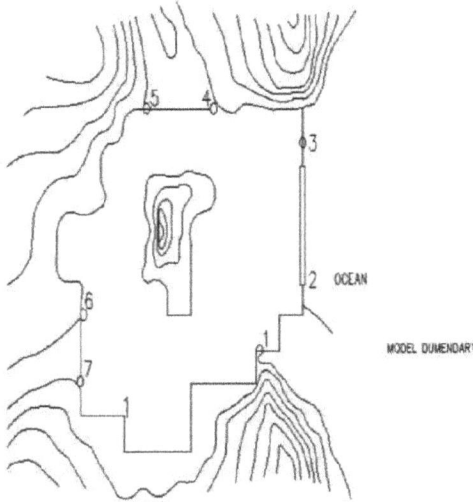

Figure 17.
A hypothetical bay.

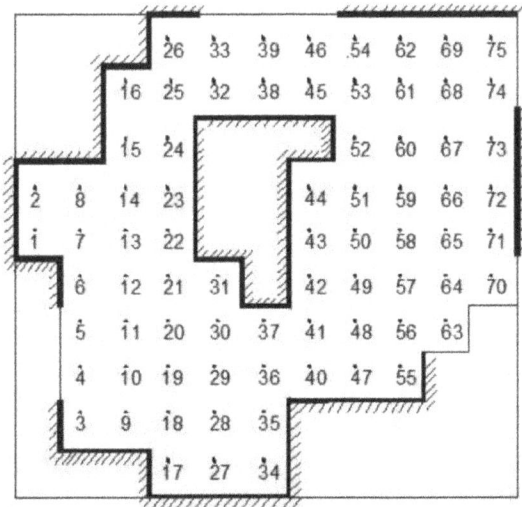

Figure 18.
The schematization of a hypothetical bay shown in **Figure 17.**

Boundary value equation: $z = a \sin\left[\frac{2\pi(t-\xi)}{T}\right] + M + 100$			
in which			
a = amplitude, and t = time (s)			
ξ = phase lag, and T = tidal period = 12.4 h = 44640 s			
M = mean water level			
NODE	a (ft)	ξ (sec)	M (ft)
63	5	0	0
70	4.95	60	0
74	4.85	180	0
75	4.85	180	0
46	4.75	1200	0.3
39	4.725	1260	0.35
33	4.7	1320	0.4
5	4.5	1800	0.7
4	4.45	1860	0.75

Table 1.
Boundary values for flow computation in a hypothetical bay.

2.5 Application 6: estuary modeling

Figure 17 illustrates a hypothetical bay, which is schematized in Figure 18. Stage hydrographs are available at seven stations as marked in Figure 17 and are numbered 1–7 (counterclockwise). Stage values in this application are expressed by sinusoidal equations (see Table 1). Some DHM-predicted flow patterns in the estuary are shown in Figures 19–21. The flow patterns appear reasonable by comparing the fluctuations of the water surface to the stage

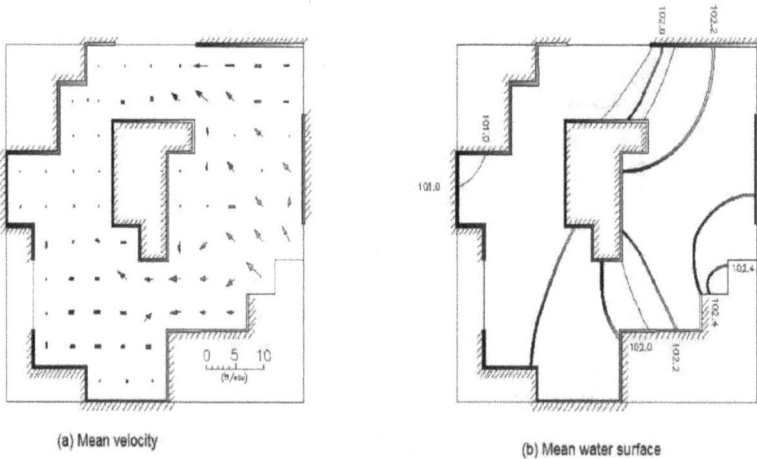

(a) Mean velocity

(b) Mean water surface

Figure 19.
Mean velocity and water surface profiles at 1 h.

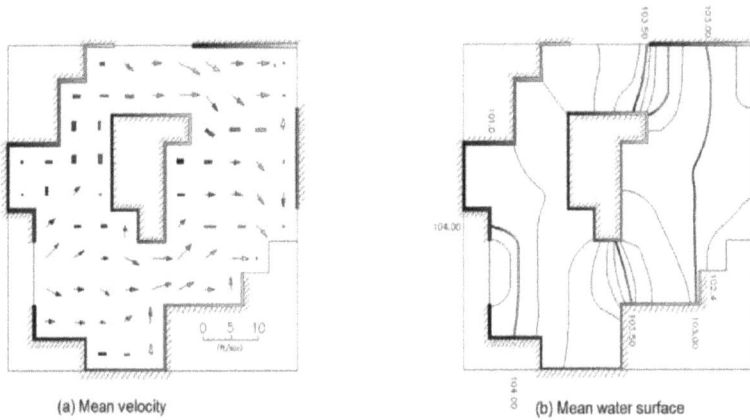

Figure 20.
Mean velocity and water surface profiles at 5 h.

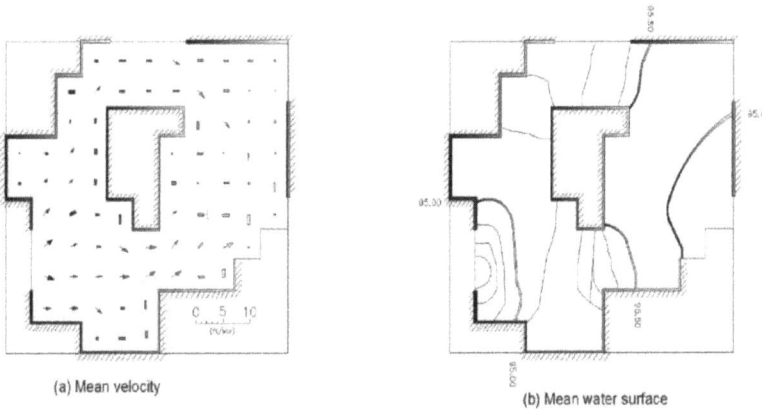

Figure 21.
Mean velocity and water surface profiles at 10 h.

hydrographs. DHM computed flow patterns compare well to a similar study prepared by Lai [5].

3. Application for channel and floodplain interface model

3.1 Application 7: channel-floodplain model

Figure 22 depicts a discretization of a two-dimensional hypothetical watershed with three major channels crossing through the flood plain.

Figure 23 depicts the inflow and outflow boundary conditions for the hypothetical watershed model. **Figures 24–30** illustrate the evolutions of the flood plain.

The shaded areas indicate which grid element is flooded. From **Figure 24**, it is seen that the outflow rates at nodes 31, 71, and 121 are less than the corresponding inflow rates, which result in a flooding situation adjacent to the outflow grid elements. The junction of channel B and B′ is also flooded. At the end of the peak

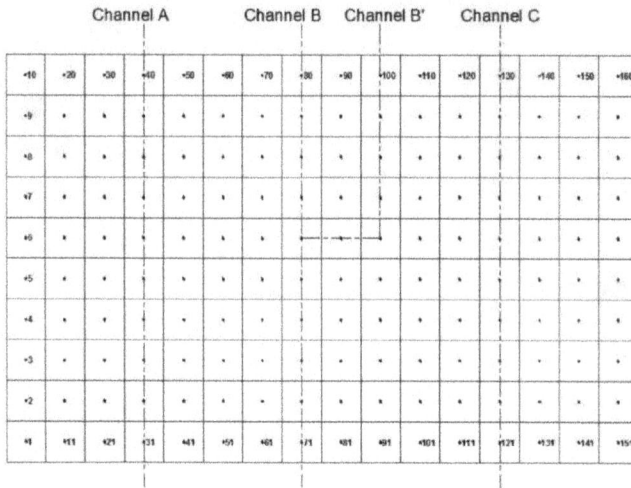

Figure 22.
Diffusion hydrodynamic model discretization of a hypothetical watershed model.

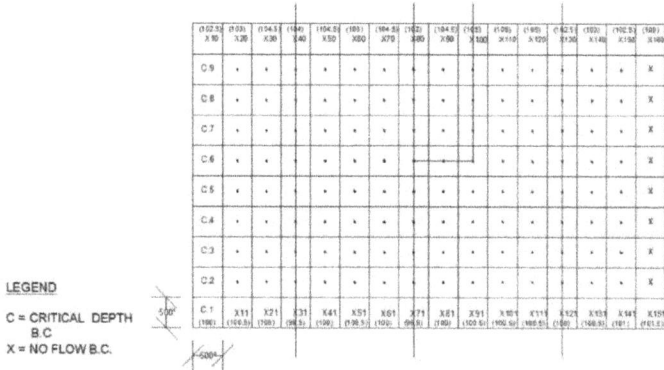

Figure 23.
Inflow and outflow boundary conditions for the hypothetical watershed model.

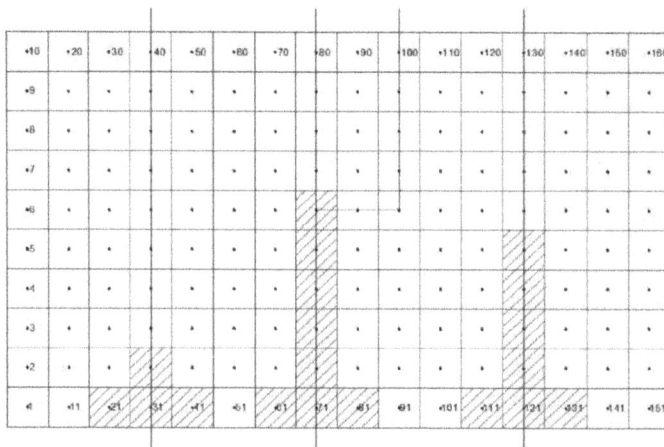

Figure 24.
Diffusion hydrodynamic modeled floodplain at time = 1 h.

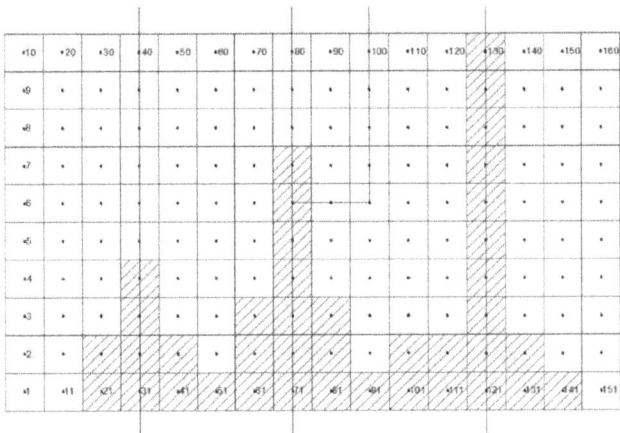

Figure 25.
Diffusion hydrodynamic modeled floodplain at time = 2 h.

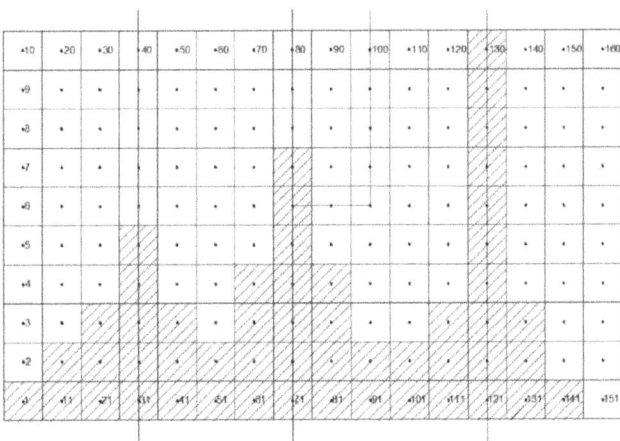

Figure 26.
Diffusion hydrodynamic modeled floodplain at time = 3 h.

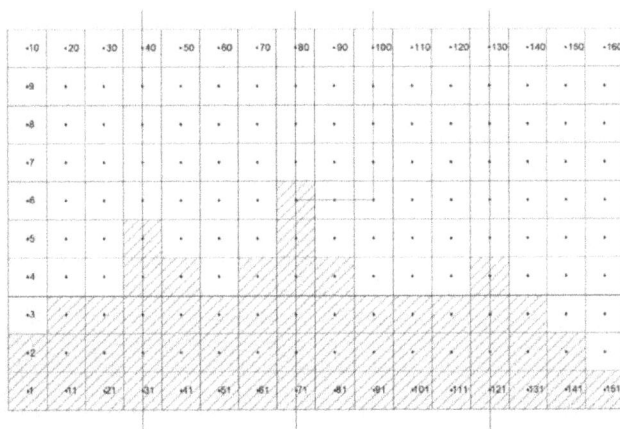

Figure 27.
Diffusion hydrodynamic modeled floodplain at time = 5 h.

inflow rate (**Figure 26**), about 1/3 of the flood plain is flooded. **Figure 29** indicates a flooding situation along the bottom of the basin after 10 h of simulation. **Figure 30** shows the maximum depth of water at four downstream cross sections. It is needed to point out that the maximum water surface for each grid element is not

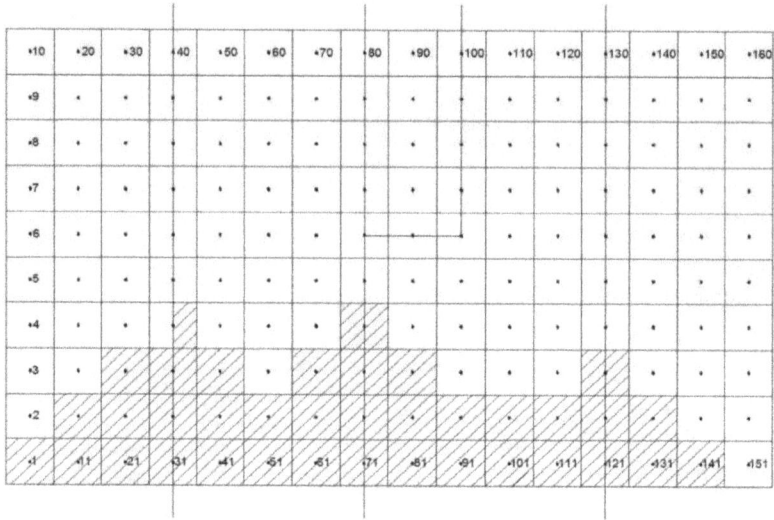

Figure 28.
Diffusion hydrodynamic modeled floodplain at time = 7 h.

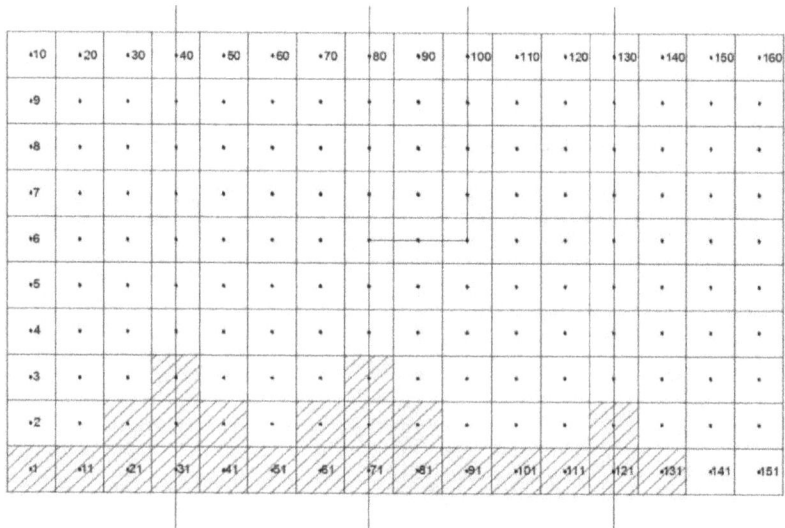

Figure 29.
Diffusion hydrodynamic modeled floodplain at time = 10 h.

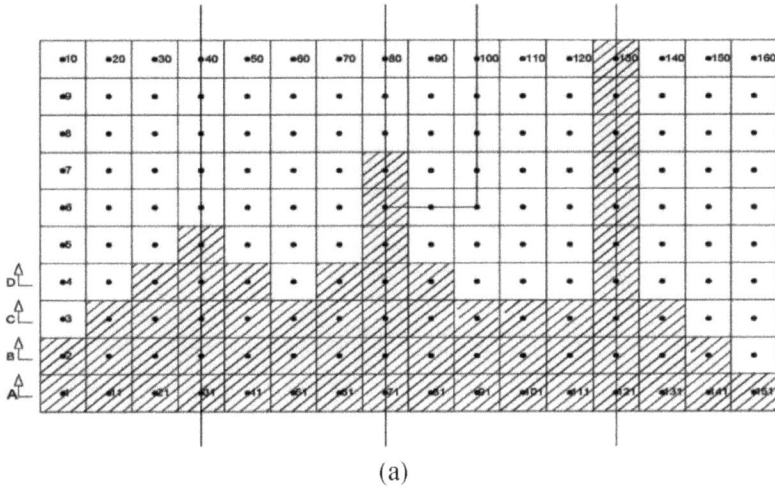

(a)

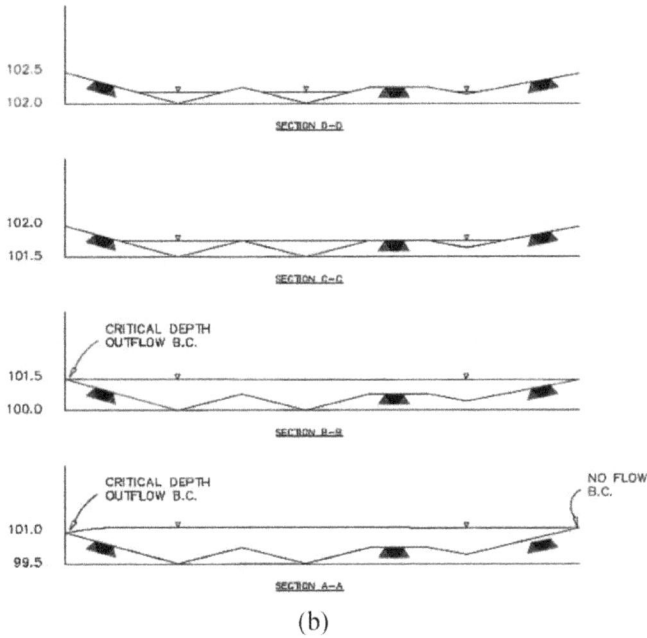

(b)

Figure 30.
Maximum water depth at different cross sections. (a) Maximum floodplain and (b) maximum water profiles.

necessarily incurred at the same time. Finally, **Figures 31** and **32** depict the outflow hydrographs for both the channel system and the flood plain system.

Until now, no existing numerical model can successfully simulate or predict the evolution of the channel-floodplain interface problem. The proposed DHM uses a simple diffusion approach and interface scheme to simulate the channel-floodplain interface development.

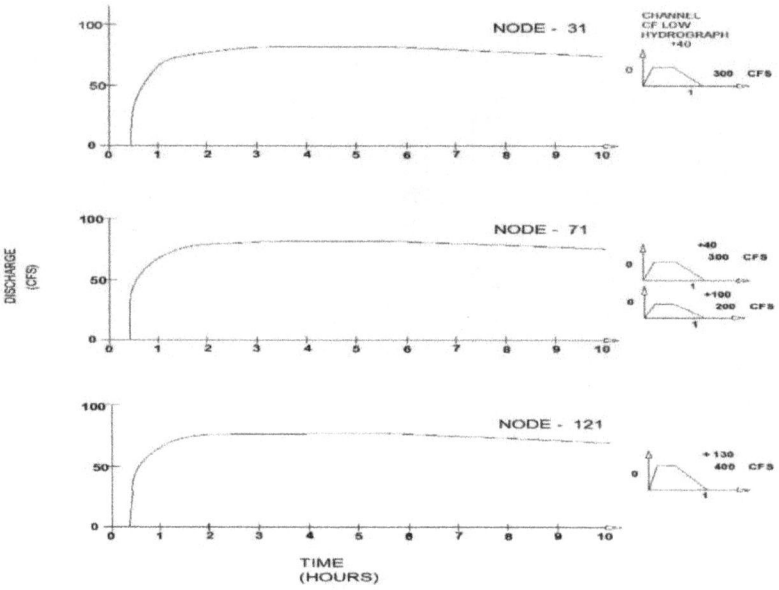

Figure 31.
Bridge flow hydrographs assumed outflow relation (Q = 10 d).

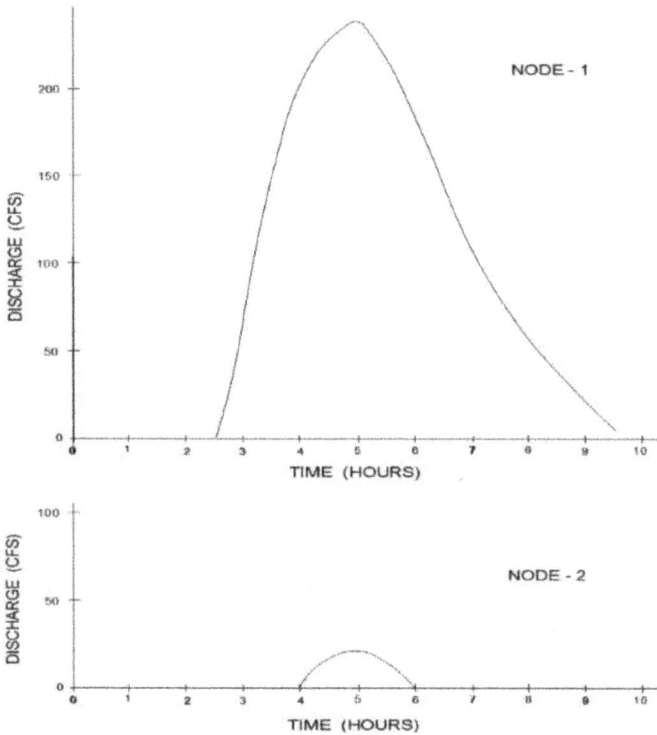

Figure 32.
Critical outflow hydrographs for floodplain.

Author details

Theodore V. Hromadka II[1*] and Chung-Cheng Yen[2]

1 Department of Mathematical Sciences, United States Military Academy, West Point, NY, USA

2 Tetra Tech, Irvine, CA, USA

*Address all correspondence to: ted@phdphdphd.com

IntechOpen

References

[1] Henderson FM. Open Channel Flow. New York: MacMillan Publishing Co., Inc.; 1966. p. 522

[2] Hromadka TV II, Nestlinger AB. Using a two-dimensional diffusional dam-break model in engineering planning. In: Proceedings: ASCE Workshop on Urban Hydrology and Stormwater Management, Los Angeles County Flood Control District Office. Los Angeles, California; 1985

[3] Hromadka TV II, Lai C. Solving the two-dimensional diffusion flow model. In: Proceedings: ASCE Hydraulics Division Specialty Conference. Orlando, Florida; 1985

[4] Metropolitan Water District of Southern California. Dam-Break Inundation Study for Orange County Reservoir. Los Angeles, California; 1973

[5] Lai C. Computer Simulation of Two-Dimensional Unsteady Flows in Estuaries and Embayments by the Method of Characteristics—Basic Theory and the Formulation of the Numerical Method. Reston, Virginia: U.S. Geological Survey, Water Resources Investigation; 1977. pp. 77-85

Chapter 5

Reduction of the Diffusion Hydrodynamic Model to Kinematic Routing

Theodore V. Hromadka II and Chung-Cheng Yen

Abstract

In this chapter, the kinematic routing option of the diffusion hydrodynamic model for one-dimensional flows is presented along with the underlying pinning of kinematic flow. The kinematic model results are compared with the full model and K-634 model output data for the mild and steep channel.

Keywords: one-dimensional flow, kinematic routing, channel slope, flow discharge, routing

1. Introduction

The two-dimensional DHM formulation as given by Eq. (32) in Chapter 1 can be simplified into a kinematic wave approximation of the two-dimensional equations of motion by using the slope of the topographic surface rather than the slope of the water surface is the friction slope in Chapter 1, Eq. (28). That is, flow rates are driven by Manning's equation, while backwater effects, reverse flows, and ponding effects are entirely ignored. As a result, the kinematic wave routing approach cannot be used for flooding situations such as considered in the previous chapter. Flows which escape from the channels cannot be modeled to pond over the surrounding land surface nor move over adverse slopes nor are backwater effects being modeled in the open channels due to constrictions which, typically, are the source of flood system deficiencies.

In a report by Doyle et al. [1], an examination of approximations of the one-dimensional flow equation was presented. The authors write: "It has been shown repeatedly in flow-routing applications that the kinematic wave approximation always predicts a steeper wave with less dispersion and attenuation than may actually occur. This can be traced to the approximations made in the development of the kinematic wave equations wherein the momentum equation is reduced to a uniform flow equation of motion that simply states the friction slope is equal to the bed slope. If the pressure term is retained in the momentum equation (diffusion wave method), then this will help to stop the accumulation of error that occurs when the kinematic wave approximation procedure is applied." More background information relating to relating to kinematic and diffusion wave equations can be found in the works of Ponce and others [2–5].

2. Application 8: kinematic routing (one-dimensional)

To demonstrate the kinematic routing feature of the DHM, the one-dimensional channel problem used for the verification of the DHM is now used to compare results between the DHM and the kinematic routing.

For the steep channel, both techniques show similar results up to 10 miles for the maximum water depth (**Figure 1**) and discharge rates at 5 and 10 miles (**Figures 2–5**). For the mild channel, the maximum water surface and discharge rates deviate increasingly as the distance increases downstream from the point of channel inflow.

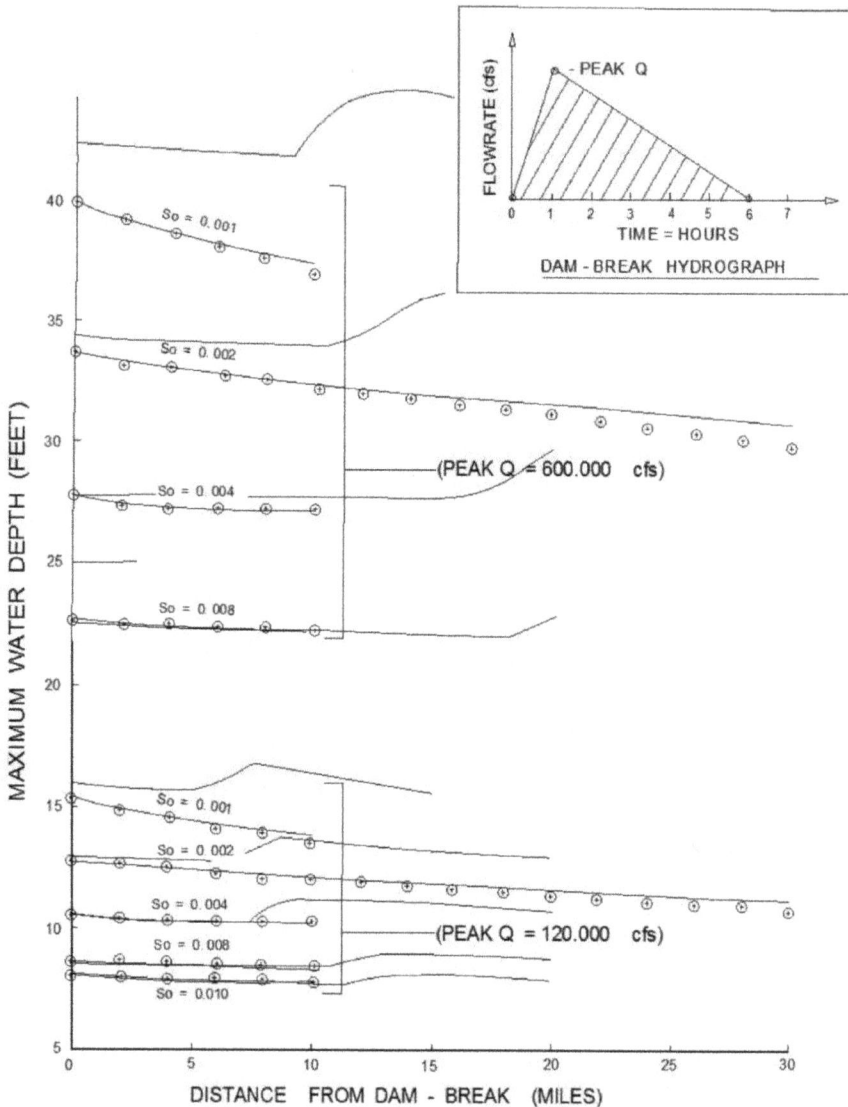

Figure 1.
Diffusion model (⊙), kinematic routing (dashed line), and K-634 model results (solid line) for 1000 feet width channel, Manning's n = 0.040, and various channel slopes, S₀.

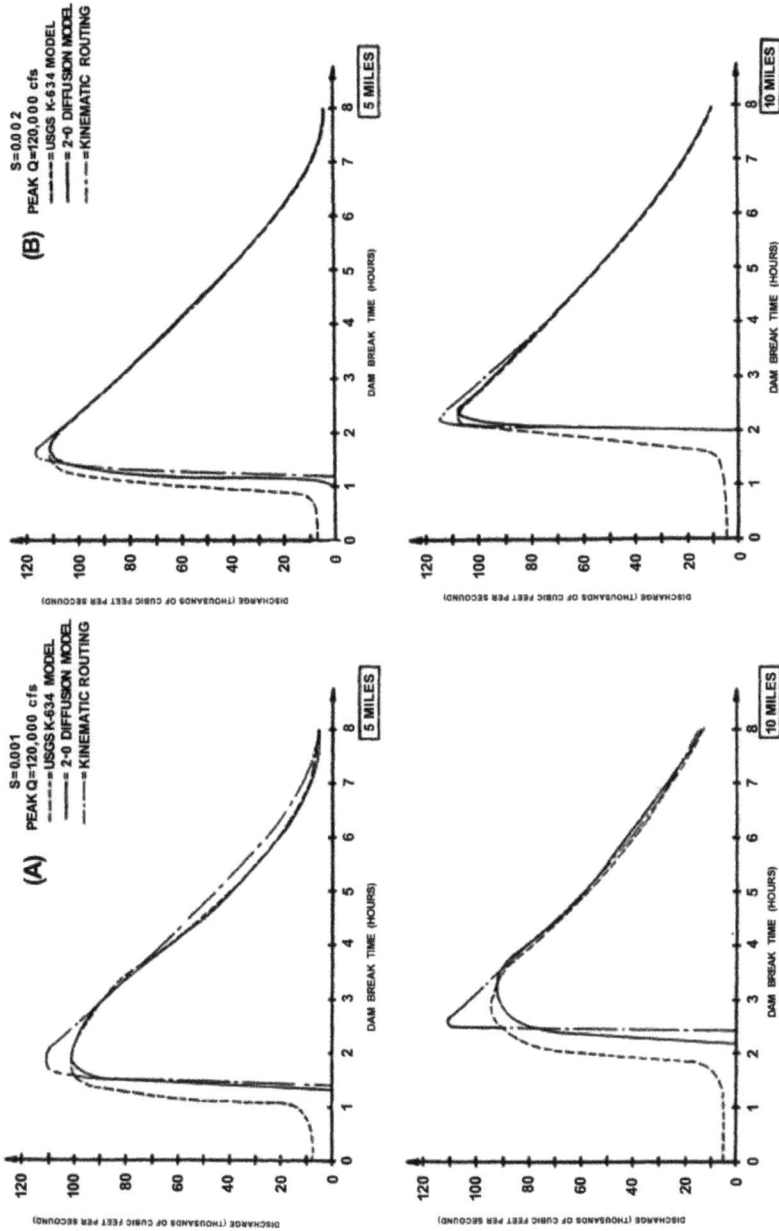

Figure 2.
Comparisons of outflow hydrographs at 5 and 10 miles downstream from the dam-break site (peak
Q = 120,000 cfs).

3. Conclusions

The reliability of the kinematic routing option in DHM was tested by applying it over mild and steep channels. The DHM results were compared with the diffusion model and K-634 model results. The close agreement of the results for the steep channel between the models across the length of flow underscores the reliability of this option in DHM.

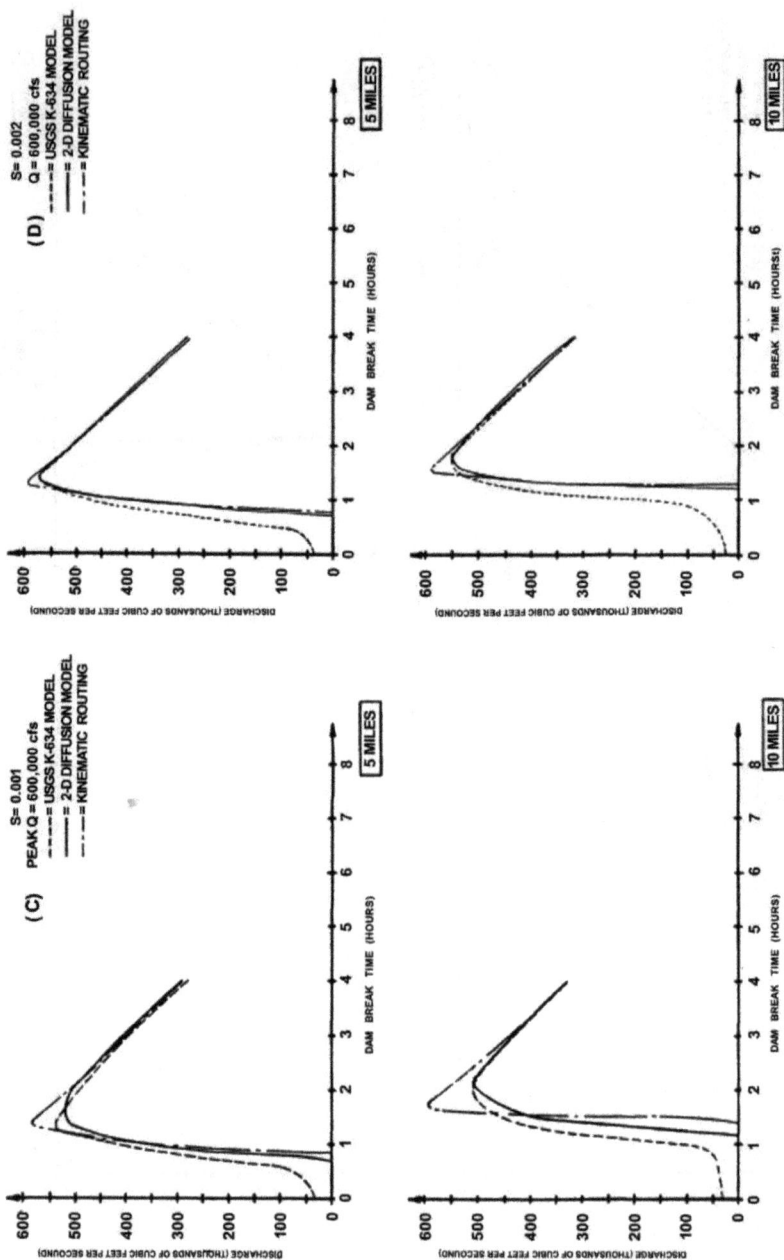

Figure 3.
Comparisons of outflow hydrographs at 5 and 10 miles downstream from the dam-break site (peak Q = 600,000 cfs).

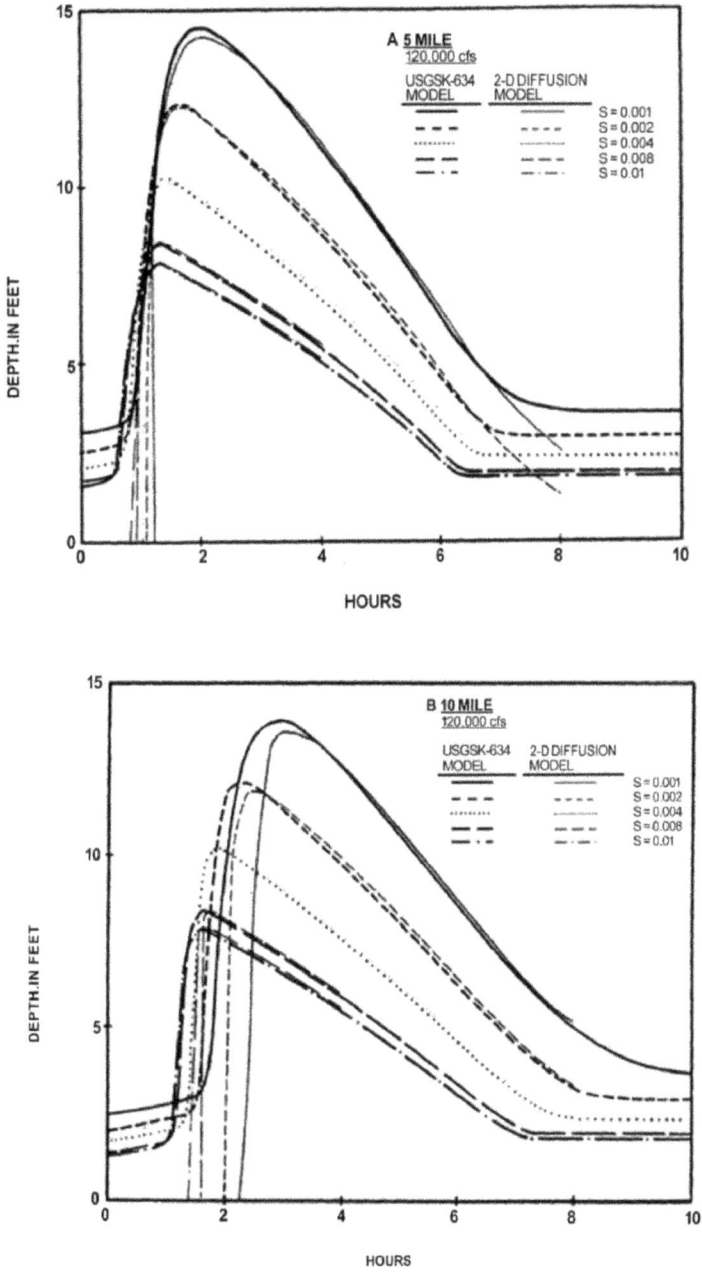

Figure 4.
Comparisons of depths of water at 5 and 10 miles downstream from the dam-break site (peak Q = 120,000 cfs).

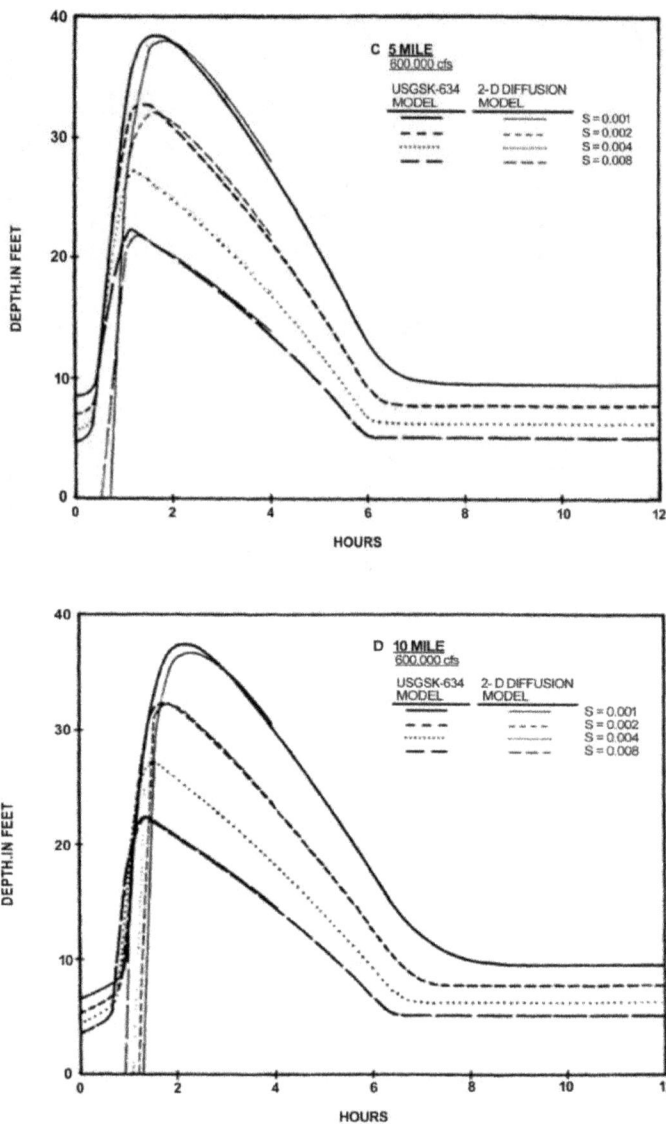

Figure 5.
Comparisons of depths of water at 5 and 10 miles downstream from the dam-break site (peak Q = 600,000 cfs).

Author details

Theodore V. Hromadka II [1*] and Chung-Cheng Yen[2]

1 Department of Mathematical Sciences, United States Military Academy,
West Point, NY, USA

2 Tetra Tech, Irvine, CA, USA

*Address all correspondence to: ted@phdphdphd.com

IntechOpen

References

[1] Doyle WH, Shearman JO, Stiltner GJ, Krug WR. A digital model for Streamflow routing by convolution method. Water-Resources Investigations Report. 1983:83-4160. DOI: 10.3133/wri834160

[2] Ponce VM. Modeling surface runoff with kinematic, diffusion, and dynamic waves. In: Singh VP, Kumar B, editors. Proceedings of the International Conference on Hydrology and Water Resources, New Delhi, India, December 1993. Dordrecht: Water Science and Technology Library, Springer; 1996. p. 16

[3] Moramarco T, Pandolfo C, Singh VP. Accuracy of kinematic wave approximation for flood routing, II: Unsteady analysis. Journal of Hydrologic Engineering. 2008;**13**(11):1089-1096

[4] Moramarco T, Pandolfo C, Singh VP. Accuracy of kinematic wave approximation for flood routing, II: Unsteady analysis. Journal of Hydrologic Engineering. 2008;**13**(11):1089-1096

[5] Xiong Y, Melching CS. Comparison of kinematic wave and nonlinear reservoir routing of urban watershed runoff. Journal of Hydrologic Engineering. 2005;**10**(1):39-49

Chapter 6

Comparison of DHM Results for One- and Two-Dimensional Flows with Experimental and Numerical Data

Theodore V. Hromadka II and Prasada Rao

Abstract

In this chapter, the performance of DHM for one- and two-dimensional flows is compared with the results of HEC-RAS, HEC-RAS 2D, WSPG, TUFLOW, Mike 21, and OpenFOAM models. The latter four models are currently widely used in industry, and benchmarking their data with DHM can shed more light on the reliability of DHM. As the results indicate, for applications which do not violate the assumptions made in DHM, the results are in agreement.

Keywords: HEC-RAS, WSPG, TUFLOW, OpenFOAM, Froude number, hydraulic jump, overland flow

1. Introduction

Numerical modeling of free surface flow across real-life applications is gaining momentum. These model domains are characterized by thousands of computational cells, and the physical characteristics have varying complexities. Over the last three decades, with the growth of computational and visualization resources, multiple numerical models have been developed for solving the free surface flow equations across one, two, and three dimensions. While some of these models are available for free in the public domain, others are licensed by their respective vendors. Based on the assumptions used in these models, the complexity of flow equations can range from Bernoulli's energy equation to three-dimensional unsteady Navier-Stokes equations. The models continue to evolve as the physics of flow is better understood, and the need for accurately predicting the flow variables across large spatial and temporal domains as their values is an important factor in the hydraulic design of structures and other related applications. Computational fluid dynamics (CFD) models that focus on solving the complete Navier-Stokes equations, rather than the energy equation or shallow water equations which are used in the hydraulic models, are also gaining popularity among the hydraulic modeling community.

The goal of this chapter is to evaluate the DHM results with a few industry-wide established software and experimental data to underscore the advantages and limitations in the models. To this end, we have chosen one critical application each from one-dimensional and two-dimensional flows.

IntechOpen

2. One-dimensional application

2.1 Flows with hydraulic jump

Modeling flows with hydraulic jump where the flow transits from super critical to subcritical has been used by different researchers [1–4] to test the reliability of their numerical formulations. Hydraulic jump is often created inflows to dissipate the flow energy, which can otherwise among others, erode the channels. They occur in gravity flows and are characterized by a large variation in flow depth and velocity before (Froude number > 1) and after (Froude number < 1) the jump. While capturing the internal flow details like bubble breakup, turbulence characteristics, tracking the water surface, aeration, fluid mixing, and turbulence is not possible using the shallow water equations, these equations can, however, predict the location and the flow depths before and after of the jump at steady state, which are important variables in the design calculations.

2.2 Experimental setup and model variables

A dataset from a series of experiments [5] that were conducted in the hydraulics laboratory at California State University, Fullerton, to simulate the location of steady-state hydraulic jump, was used for validating the models. The rectangular open channel flume was 15.2 m long, 0.46 m wide and 0.6 m in height. The channel sides are of glass, while the bottom interface with water is a metal sheet with a Manning roughness coefficient of 0.01. The bottom slope of the channel can be changed by tilting the flume, and in this investigation, it was set to 0.012. The flow discharge is 0.036 m^3/s.

Boundary conditions need to be consistent with the physics of flow and appropriately complement the flow Equations [6]. The number of boundary conditions at the two ends of the flow domain is governed by the local Froude number. From a mathematical perspective, a boundary condition is a constraint imposed at the boundary node to arrive at a unique solution to a well-posed equation set. Specifying more or less than the required number may make the problem "ill-posed" and can lead to incorrect solutions. While one-dimensional supercritical flow requires superimposing two boundary conditions at the upstream end, a subcritical flow requires superimposing one boundary condition at the downstream end. In this simulation, at the upstream end, a flow depth of 0.04 m and flow discharge was specified. At the downstream end, a constant flow depth of 0.24 m was used. For this flow and boundary depth combination, the Froude numbers at the upstream and downstream end of the channel are 3.37 and 0.66, respectively.

2.3 Examined numerical models

The results of the DHM, RAS, WSPG, and TUFLOW models were compared with the experimental data. The other three models are briefly described below.

HEC-River Analysis System (RAS): HEC-RAS (steady state) model is based on the solution of the one-dimensional energy equation between two sections with energy losses given by Manning's equation. The momentum equation can be used in situations where the water surface profile is rapidly varied as in hydraulic jump, hydraulics of bridges, and evaluating profiles at river confluences (stream junctions). RAS model also has modules to solve unsteady flows, sediment transport, and water quality analysis. In this work, the steady-state model was used. The model was developed by the US Army Corps of Engineers and can be downloaded for free [7].

Water Surface Pressure Gradient (WSPG): WSPG is one of the first models in computational hydraulics that was developed by the Los Angeles County Department of Public Works. The model solves the Bernoulli energy equation between any two cross sections, using the standard step method. The program computes uniform and nonuniform steady flow water surface profiles. As part of the solution, it can automatically identify a hydraulic jump in the channel reach. The model is currently distributed for a fee by civil design [8].

Two-dimensional unsteady flow (TUFLOW): The two-dimensional depth-averaged shallow water equations are solved in TUFLOW using a structured grid system with an alternating direction implicit scheme. The algorithm can capture flow transitions from supercritical to subcritical. TUFLOW incorporates the 1-D component (ESTRY software) or quasi-2D modeling system based on the full one-dimensional free surface flow Equations [9]. The model was developed by BMT WBM and can be downloaded for a fee.

2.4 Results

Figure 1 is a plot of the steady-state depth profile from the four models together with the experimental data. But for DHM, all other models satisfactorily predict the location and the flow depth before and after the jump. The reason as to why DHM could not capture the jump is because of the number of boundary conditions that the DHM permits from the end user. At the upstream end, DHM allows for only the flow discharge to be specified (and not two boundary conditions). The model had 18 grid elements in the computational domain. The upstream element is #1, and the end downstream element is #18. Element #14 corresponds to the end of the channel (length = 15.2 m). At element #1, the input discharge is 0.036 m^3/s. At element #18, critical depth condition is specified, and the grid elevation was progressively raised such the depth at element #14 equals 0.24 m.

Because of this boundary condition limitation in the DHM, the jump is smoothened out in the solution. Although the downstream depth is consistent with other models, at the upstream end, the DHM predicted depth is higher than actual depth. DHM computed the flow transitioning from supercritical to subcritical without going through a hydraulic jump as required by theory and observed in the flume

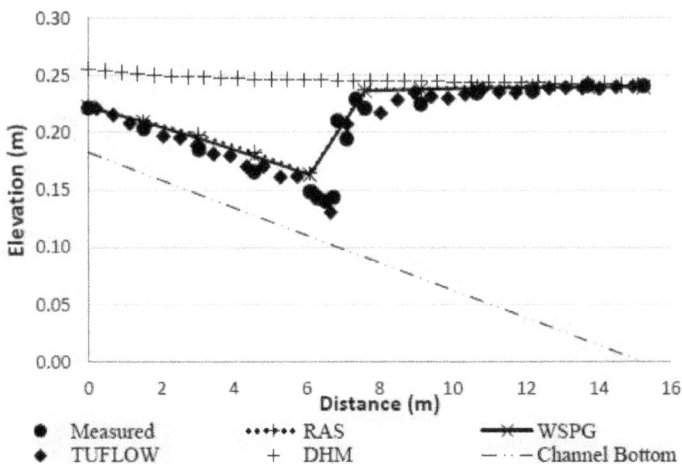

Figure 1.
Steady-state results for one-dimensional rapidly varying flow.

model. It can be concluded that DHM cannot be used in applications which require the prediction of the location of hydraulic jump.

3. Two-dimensional applications

3.1 Overland flow on a sloping domain

Overland flow is a dynamic response of the watershed to excess rainfall. Overland flow typically occurs as sheet flow on the land surface, and when the flow joins a channel, it is known as streamflow. The spatial and temporal distribution of two-dimensional overland flow variables is driven by the topographical characteristics of the domain and the boundary conditions. The nonhomogeneous surface characteristics and varying width of the natural watercourse path in the direction of flow make it an ideal case for comparing the results of different models.

Modeling overland flow has drawn the attention of many researchers. Since excess rainfall can cause flooding and mudslides which has the potential to cause loss of human life and disrupt the local economy, reliably predicting the flow variables for different precipitation scenarios can assist decision-makers and emergency personnel. Researchers have modeled these flows by solving the two-dimensional fully dynamic shall water Equations [10, 11] or their diffusion [12, 13] or kinematic approximations [14, 15].

3.1.1 Examined numerical models

The results of the DHM, extended DHM (EDHM), RAS, WSPG, and the CFD (OpenFOAM) models were compared. The back engine for EDHM is identical to DHM, with the primary difference between the two models being the larger array size of the variables in EDHM. When DHM was originally developed, the maximum array dimension was limited to 250, largely because of the available computational resources in 1980s. In EDHM, the dimension of all the arrays was increased to 9999. Since background information relating to RAS and WSPG has been given earlier, characteristic features of the CFD model, OpenFOAM, are summarized below.

OpenFOAM (Open-source Field Operation and Manipulation) is a freely available open-source software [16] that is gaining popularity across CFD applications. Its versatile C++ toolbox for the Linux operating system enables developing customized, efficient numerical solvers, and pre-/post-processing utilities for all kinds of CFD flows [17]. This code uses a tensorial approach following the widely known finite volume method (FVM), first used by McDonald [18]. Both structured and unstructured meshes can be used in the computational domain. The time integration can be done through backward Euler, steady-state solver, and Crank-Nicholson. The available gradient, divergence, Laplacian, and interpolation schemes are second-order central difference, fourth-order central difference, first-order upwind, and first-/second-order upwind. The turbulence models that can be used in OpenFOAM are LES, k-ε, and k-ω. The available solvers, options in specifying the boundary conditions, mesh generation tools, flow visualization software, and extensive documentation are making OpenFOAM popular among the CFD modeling community [2, 13, 19].

3.1.2 Study area and model variables

The study area is shown in **Figure 2**. Overland flow generated by a storm down a steep slope hits the flat main street after which a significant portion of the flow continues flowing North. The flow is supercritical from the inlet boundary location to the downstream of the main street. At the street downstream, there is a wall which reduces the flow velocity, thus forcing the flow to be critical. In this analysis, lateral flow on the main street was neglected.

In the models, the Manning roughness coefficient was set to 0.015 for the street portion and 0.03 for the earth. A uniform grid size of 15 ft was used, which resulted in a total of grids in the domain. The grids were oriented with the natural watercourse (NWC) path. The NWC ranged between 27 and 35 ft in the vicinity of the upstream end (southwest corner), and its width ranged between 45 and 60 ft in the vicinity where the water hits the main street. Having a 15 ft grid enabled us to cover the entire NWC path (**Figure 3**). The elevation at the center of the grid in the DHM was obtained from the topography map of the area. For the RAS and WSPG models, the required cross-sectional data was obtained from the top map. The rest of the input variables were consistent with the DHM data.

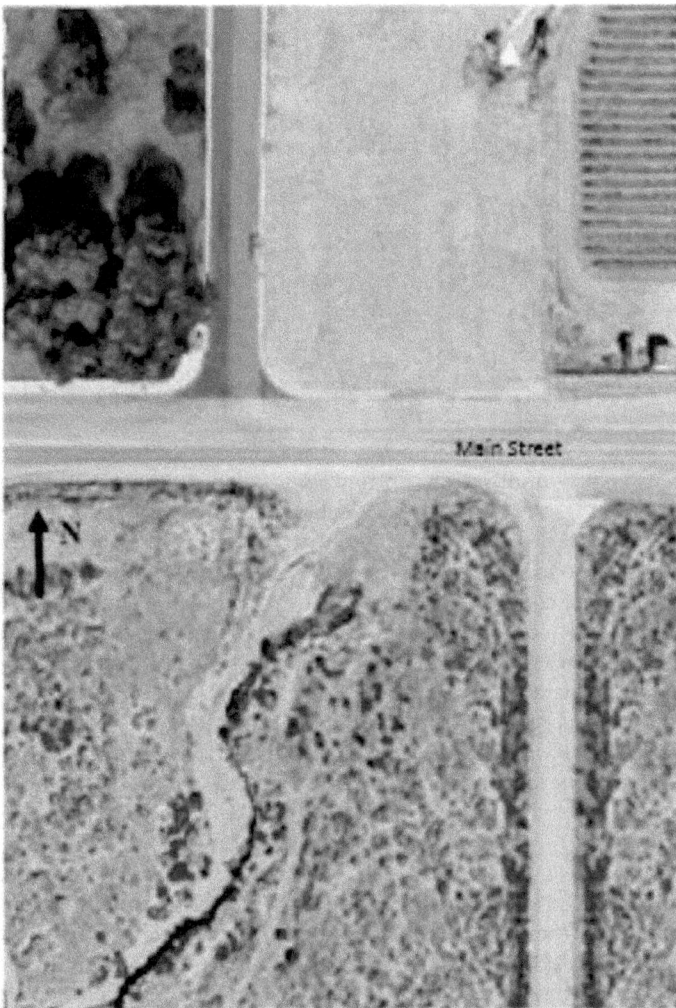

Figure 2.
Map of the study area. The inflow and outflow boundary locations are identified by ▲ and   .

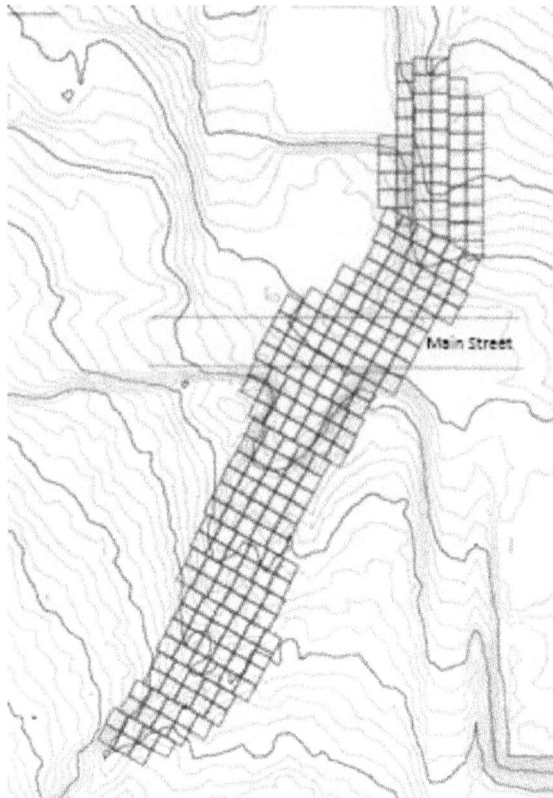

Figure 3.
DHM computational domain. The domain is aligned with the natural watercourse path and had 248 cells, which are 15 ft squares.

The intensity of rainfall and the bottom slope enhances the power of gravity-driven overland flow to make it supercritical (Froude number > 1). The available power in the water near the street can potentially push any moving or stationary automobiles. At the upstream end center cell, the flow hydrograph (**Figure 4**) was used as the boundary condition. The peak discharge at t = 0.5 hours is 755 cfs. At the downstream end, a critical depth boundary was specified. To keep the effect of downstream boundary minimal on the solution, the domain was extended by about 250 ft north of the main street.

Our focus was on predicting the flow depth at multiple probe locations on the main street. To conserve space, these results are plotted at two of the thirteen probe locations. These probe locations are 9 and 13 (**Figure 5**).

3.1.3 Results

The DHM computational results include a variety of hydraulics relationships that are useful in further detailed analysis, such as flow velocity, flow depth, Froude number, and so forth. Of course, the code can be readily included in the DHM or as a post-processor routine, which enhances the DHM outcome. Of particular interest are the computational results from the DHM in comparison with the computational results produced by the CFD application. To display these computational results, hypothetical "probes" are inserted into the computational mesh where

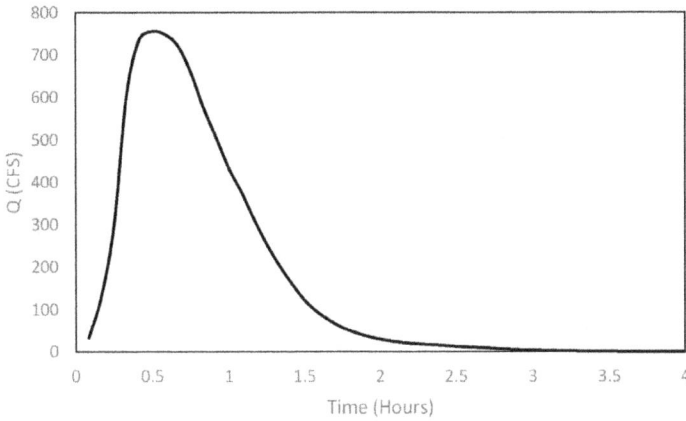

Figure 4.
Input hydrograph at the upstream end.

Figure 5.
Location of the probes along the center of the main street. The flow depth results were compared at probe locations 9 and 13.

computational results are assembled and collated into a form suitable for visualization. The results from the visualization assessment are depicted below.

Figures 6 and 7 are plots comparing the flow depth, at probe location 9 and 13 on the main street. The data from DHM, EDHM, WSPG, RAS, and CFD (OpenFOAM) are plotted, for a specific CFD simulation time period. It is noted that the computational results are of high similarity, yet the computational effort ranges vastly. The DHM takes approximately 1 hour of CPU for the indicated simulation.

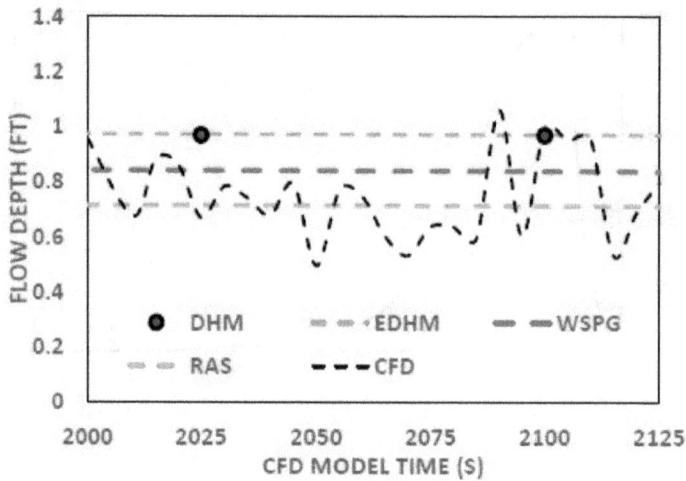

Figure 6.
Comparison of flow depth at probe 9 location.

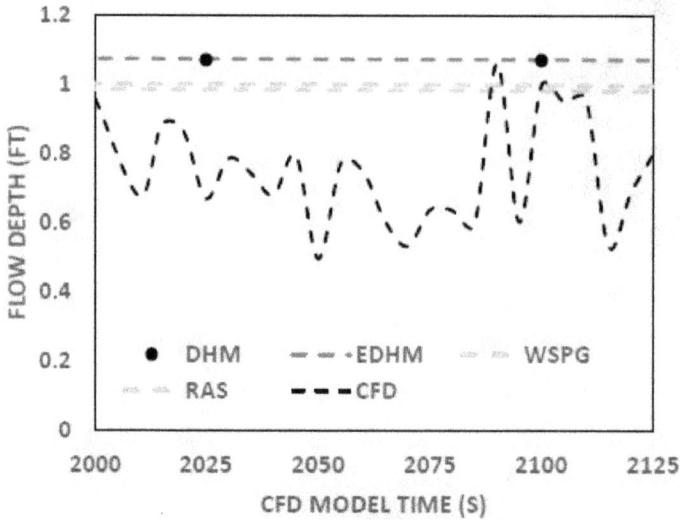

Figure 7.
Comparison of flow depth at probe 13 location.

In comparison, because of the varying small grid sizes in the domain, the CFD model required 2 weeks of CPU time using a parallel processor.

3.2 Open channel flow with a constriction

Numerically predicting the characteristics of flow through a channel with symmetric abrupt constriction (**Figure 8**) in the form of reduced channel width has drawn the attention of many researchers and has been part of any standard text book in hydraulics. The idea of developing DHM for this application was inspired after reading a recent paper [20] who tested a series of 2D models for multiple applications, one of which is flow through a constriction. Our focus was to estimate the DHM head loss at steady state and compare it with the published data.

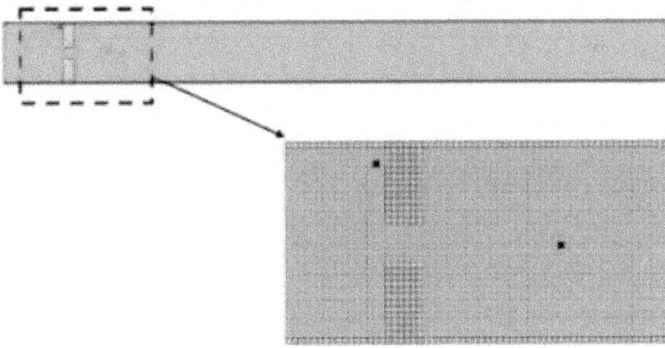

Figure 8.
Definition sketch of the test problem along with the location of the two points (P1 and P2).

3.2.1 Examined numerical models

The results of the extended DHM (EDHM), Mike 21, TUFLOW, and HEC-RAS 2D models were compared, with the bench mark data from the equations provided by the Federal Highways Administration (FHA). Mike 21 and HEC-RAS 2D are briefly outlined here. Mike 21 solves the two-dimensional free surface flows where stratification can be neglected. It was originally developed for flow simulation in coastal areas, estuaries, and seas. The various modules of the system simulate hydrodynamics, advection-dispersion, short waves, sediment transport, water quality, eutrophication, and heavy metals [21].

HEC-RAS 2D (5.0.1) solves the two-dimensional Saint-Venant Equations [7] for shallow water flows using the full momentum computational method. The equations can model turbulence and Coriolis effects. For flow in sudden contraction, which is accompanied by high velocity, using the full momentum method in RAS 2D is recommended. The model uses an implicit finite volume solver.

3.2.2 Model variables

Figure 8 is the definition sketch of the test problem. The rectangular channel is 3100 ft long and 320 ft wide. The constriction is 60 × 60 ft. The channel length before constriction is 310 ft, and its length after constriction is 2730 ft. The computational domain in EDHM had 10 ft square cells, and the total number of cells was 9920. The longitudinal slope is 1%, the transverse slope is zero, and the model was run for a total of 1 hour. The upstream inflow is 1000 cfs. Since there are 30 cells at the upstream end, a uniform steady inflow of 33.3 cfs was specified at each of the cells. At the downstream end, a free overall boundary was specified. Constricting the flow area results in loss of energy. This loss of energy is reflected in a rise in energy gradient line and energy upstream of the constriction. Of interest is to estimate the head loss that occurs between points 1 and 2 (shown in **Figure 8**). The head loss (H_L) equals $WSE_2 - WSE_1$, where WSE is the water surface elevation [20].

3.2.3 Results

Table 1 shows the comparison of the head loss value obtained from the models along with the published data of other models. It is noted that in this effort, the computational models are compared with respect to head loss (as given in [20]) through the constriction, and this is the primary form of assessment. The DHM

	WSE change (ft)
HEC-RAS 2D	1.27
TUFLOW	0.99
Mike 21	1.28
EDHM	1.16
FHA equation	0.8

*Except for EDHM all other data were obtained from the literature [20].

Table 1.
Comparison of change in water surface elevation at constriction between EDHM and published data.

WSE change value is within the range of other model predictions, although all the model predictions are above the FHA value.

4. Conclusions

Results from multiple computer models are compared with those of DHM for one- and two-dimensional flows. The considered one-dimensional flow was a mixed flow with a hydraulic jump. All the model results (DHM, WSPG, RAS, TUFLOW) were compared with the benchmark experimental data. Because of the way the boundary conditions are specified in the DHM, the model cannot simulate the hydraulic jump. For the two-dimensional overland flow, the model results (DHM, EDHM, TUFLOW, MIKE 21, WSPG, RAS, RAS2D, and the CFD model, OpenFOAM) were compared between themselves. The agreement of the predicted flow variables reinforces the reliability of the current model.

Author details

Theodore V. Hromadka II [1*] and Prasada Rao[2]

1 Department of Mathematical Sciences, United States Military Academy, West Point, NY, USA

2 Department of Civil and Environmental Engineering, California State University, Fullerton, CA, USA

*Address all correspondence to: ted@phdphdphd.com

IntechOpen

References

[1] Viti N, Valero D, Gualtieri C. Numerical simulation of hydraulic jumps. Part 2: Recent results and future outlook. Water. 2018;**11**:28. DOI: 10.3390/w11010028

[2] Bayón Barrachina A, López Jiménez PA. Numerical analysis of hydraulic jumps using OpenFOAM. Journal of Hydroinformatics. 2015;**17**(4):662-678. DOI: 10.2166/hydro.2015.041

[3] Mortazavi M, Le Chenadec V, Moin P, Mani A. Direct numerical simulation of a turbulent hydraulic jump: Turbulence statistics and air entrainment. Journal of Fluid Mechanics. 2016;**797**:60-94

[4] Gharangik AM, Chaudhry MH. Numerical simulation of hydraulic jump. Journal of Hydraulic Engineering ASCE. 1991;**117**:1195

[5] Rao P, Hromadka TV II. Numerical modeling of rapidly varying flows using HEC-RAS and WSPG models. Technical Note. Springerplus. 2016;**5**:662. DOI: 10.1186/s40064-016-2199-0

[6] Hirsh C. Numerical Computation of Internal and External Flows. New York (NY): John Wiley & Sons; 1990

[7] U.S. Army Corps of Engineers (USACE). HEC-RAS River Analysis System. User's Manual. Davis, CA: Hydrologic Engineering Center; 2019. Version 5.0.7. Available from: http://www.hec.usace.army.mil/software/hec-ras/

[8] Water Surface Pressure Gradient for Windows. Joseph E. Bonadiman & Associates, Inc.; Available from: https://civildesign.com/products/wspgw-water-surface-pressure-gradient-for-windows

[9] BMT-WBM, Australia. User's Manual for TUFLOW. Spring Hill: WBM Oceanics Australia; 2019

[10] Costabile P, Costanzo C, Macchione F, Mercogliano P. Two-dimensional model for overland flow simulations: A case study. European Water. 2012;**38**:13-23

[11] Ponce VM. Modeling surface runoff with kinematic, diffusion, and dynamic waves. In: Singh VP, Kumar B, editors. Proceedings of the International Conference on Hydrology and Water Resources. New Delhi, India, December 1993. Dordrecht: Water Science and Technology Library, Springer; 1996. p. 16

[12] Kazezyılmaz-Alhan CM. An improved solution for diffusion waves to overland flow. Applied Mathematical Modelling. 2012;**36**:465-1472

[13] Santillana M. Analysis and numerical simulation of the diffusive wave approximation of the shallow water equations [PhD thesis]. Austin: University of Texas; 2008

[14] Liu Q, Chen L, Li J, Singh V. Two-dimensional kinematic wave model of overland-flow. Journal of Hydrology. 2004;**291**(10):28-41

[15] Tsai T, Yang J. Kinematic wave modelling of overland flow using characteristic method with cubic-spline interpolation. Advances in Water Resources. 2005;**28**(7):661-670

[16] The Free Software Foundation Inc. OpenFOAM: The Open Source CFD Toolbox User Guide. London, United Kingdom: The Free Software Foundation Inc; 2019. Available from: https://openfoam.org/

[17] Weller H, Tabor G, Jasak H, Fureby C. A tensorial approach to computational continuum mechanics using object-oriented techniques. Computers in Physics. 1998;**12**:620-631

[18] McDonald PW. The computation of transonic flow through two-dimensional gas turbine cascades. American Society of Mechanical Engineers. Paper No: 71-GT-89, V001T01A089; 1971. p. 7. DOI: 10.1115/71-GT-89

[19] Bayon A, Valero D, García-Bartual R, Jos F, Valles-Mor FJ, Lopez-Jim P. Performance assessment of OpenFOAM and FLOW-3D in the numerical modeling of a low Reynolds number hydraulic jump. Environmental Modelling & Software. 2016;**80**:322-335

[20] Paudel M, Roman SB, Pritchard J. A Comparative Study of HEC-RAS 2D, TUFLOW, & Mike 21. In: ASFPM 2016 Annual National Conference, Grand Rapids, MI. 2016

[21] DHI, MIKE 11 & MIKE 21 Flow Model. Scientific Documentations. 2019. Available from: https://www. mikepoweredbydhi.com

www.ingramcontent.com/pod-product-compliance
Lightning Source LLC
Chambersburg PA
CBHW070242230326

41458CB00100B/5842